Oracle Certified Associate, Java SE 7 Programmer Study Guide

Comprehensive review of Oracle Certified Associate, Java SE 7 Programmer Certification objectives

Richard M. Reese

BIRMINGHAM - MUMBAI

Oracle Certified Associate, Java SE 7 Programmer Study Guide

Copyright © 2012 Packt Publishing

All rights reserved. No part of this book may be reproduced, stored in a retrieval system, or transmitted in any form or by any means, without the prior written permission of the publisher, except in the case of brief quotations embedded in critical articles or reviews.

Every effort has been made in the preparation of this book to ensure the accuracy of the information presented. However, the information contained in this book is sold without warranty, either express or implied. Neither the author, nor Packt Publishing, and its dealers and distributors will be held liable for any damages caused or alleged to be caused directly or indirectly by this book.

Packt Publishing has endeavored to provide trademark information about all of the companies and products mentioned in this book by the appropriate use of capitals. However, Packt Publishing cannot guarantee the accuracy of this information.

First published: August 2012

Production Reference: 1160812

Published by Packt Publishing Ltd.
Livery Place
35 Livery Street
Birmingham B3 2PB, UK.

ISBN 978-1-84968-732-4

www.packtpub.com

Cover Image by Mark Holland (m.j.g.holland@bham.ac.uk)

Credits

Author
Richard M. Reese

Reviewers
Edward E. Griebel Jr.
Abraham Tehrani

Acquisition Editor
Dhwani Devater

Lead Technical Editor
Susmita Panda

Technical Editor
Vrinda Amberkar

Project Coordinator
Joel Goveya

Proofreader
Linda Morris

Indexer
Hemangini Bari

Graphics
Manu Joseph

Production Coordinator
Nilesh R. Mohite

Cover Work
Nilesh R. Mohite

About the Author

Richard Reese is an Associate Professor for Computer Science at Tarleton State University in Stephenville, Texas. Previously, he has worked in the Aerospace and Telephony industries for over 17 years. He earned his Ph.D. in Computer Science from Texas A&M University and served 4 years in the Air Force primarily in the field of Communication Intelligence.

Outside the classroom, he enjoys tending his vegetable garden, maintaining his aquariums and greenhouse, and playing with his dog, Zoey. He also enjoys reading a variety of technical and non-technical material.

Richard has written numerous publications and wrote *EJB 3.1 Cookbook* and the *Java 7 New Features Cookbook* for Packt Publishing.

> No book can be written without the help of others. To this end I am thankful for the support of Karla, my wife, whose patience and reviews have made this effort possible. In addition, I would like to thank the editorial staff of Packt and the reviewers for their input which has resulted in a much better book than it might otherwise have been.

About the Reviewers

Edward E. Griebel Jr. was first introduced to computers in elementary school through LOGO on an Apple][and "The Oregon Trail" on a VAX. Pursuing his interest in computers, Ed graduated from Bucknell University with a degree in Computer Engineering. At his first job he quickly realized he didn't know everything that there was to know about Computer Programming. Ed has spent the past 20 years honing his skills in the securities trading, telecommunications, payroll processing, and machine-to-machine communications industries as a Developer, Team Lead, Consultant, and Mentor. Currently working on Enterprise Development in Java EE, Ed feels that any day spent writing code is a good day.

Away from the keyboard, Ed enjoys road trips with his wife and three children, and playing Ultimate Frisbee and Xbox. He also volunteers as his sons' scout leader and on committees at his church.

> I would like to thank my wife and three children who are by now used to letting me sleep late after long nights at the computer.

Abraham Tehrani has over a decade of experience in Software Development as a Developer and QA Engineer. He is passionate about quality and technology.

> I would like to thank my fiancé for her support and love. I would also like to thank my friends and family for supporting me in all of my endeavors.

www.PacktPub.com

Support files, eBooks, discount offers and more

You might want to visit www.PacktPub.com for support files and downloads related to your book.

Did you know that Packt offers eBook versions of every book published, with PDF and ePub files available? You can upgrade to the eBook version at www.PacktPub.com and as a print book customer, you are entitled to a discount on the eBook copy. Get in touch with us at service@packtpub.com for more details.

At www.PacktPub.com, you can also read a collection of free technical articles, sign up for a range of free newsletters and receive exclusive discounts and offers on Packt books and eBooks.

http://PacktLib.PacktPub.com

Do you need instant solutions to your IT questions? PacktLib is Packt's online digital book library. Here, you can access, read and search across Packt's entire library of books.

Why Subscribe?
- Fully searchable across every book published by Packt
- Copy and paste, print and bookmark content
- On demand and accessible via web browser

Free Access for Packt account holders

If you have an account with Packt at www.PacktPub.com, you can use this to access PacktLib today and view nine entirely free books. Simply use your login credentials for immediate access.

Instant Updates on New Packt Books

Get notified! Find out when new books are published by following @PacktEnterprise on Twitter, or the *Packt Enterprise* Facebook page.

Table of Contents

Preface	**1**
Chapter 1: Getting Started with Java	**7**
Understanding Java as a technology	**7**
Object-oriented software development	8
OOP principles	10
Examining the types of Java applications	12
Exploring the structure of a Java console program	**15**
A simple Java application	15
Package	17
Import	17
The Customer class	18
Instance variables	18
Methods	18
The CustomerDriver class' main method	19
Exploring the structure of a class	**19**
Classes, interfaces, and objects	20
Classes and objects	20
Constructors	21
Interfaces	21
Methods	22
Method declaration	22
Method signature	23
The main method	24
Access modifiers	25
Documentation	26
Comments	27
Java naming conventions	28
Using Javadocs	28
Investigating the Java application development process	**29**
Compiling a Java application	30
SDK file structure	31

Table of Contents

IDE file structure	31
Developing Java applications without an IDE	32
Java environment	33
Annotations	35
Java class libraries	35
Summary	**36**
Certification objectives covered	**37**
Test your knowledge	**37**
Chapter 2: Java Data Types and Their Usage	**39**
Understanding how Java handles data	**40**
Java identifiers, objects, and memory	40
Stack and heap	40
Declaring a variable	45
Primitive data types	46
Wrapper classes and autoboxing	48
Initializing identifiers	49
Java constants, literals, and enumerations	51
Literals	51
Constants	56
The final keyword	57
Enumerations	57
Immutable objects	58
Instance versus static data	58
Scope and lifetime	58
Scoping rules	59
Access modifiers	60
Data summary	61
Building expressions using operands and operators	**61**
Precedence and associativity	62
Casting	63
Working with characters and strings	**64**
The String, StringBuffer, and StringBuilder classes	64
Unicode characters	65
The Character class	66
The Character class – methods	66
The String class	66
String comparisons	68
Basic string methods	72
String length	73
Number/string conversions	73
Miscellaneous String methods	74
The StringBuffer and StringBuilder classes	74
Summary	**75**

Certification objectives covered	**76**
Test your knowledge	**76**
Chapter 3: Decision Constructs	**79**
Flow of control	**80**
Control statement – an overview	80
Logical expressions	**81**
Boolean variables	81
The equality operator	82
Relational operators	82
Logical operators	83
Short circuit evaluation	85
Using the && operator	85
Using the || operator	86
Avoiding short circuit evaluation	86
The if statement	**87**
Nested if statements	90
The else-if variation	91
The if statement – usage issues	92
Misusing the equality operator	92
Using inverse operations	93
Using Boolean variables instead of logical expressions	94
Using true or false in a logical expression	95
The perils of not using the block statement	96
The dangling else problem	97
Conditional operator	**99**
The switch statement	**100**
Integer-based switch statements	101
Enumeration-based switch statements	103
String-based switch statements	104
String issues with the switch statement	105
Control structure issues	**105**
General decision constructs issues	105
Floating point number considerations	106
Special floating point values	107
Comparing floating point numbers	109
Rounding errors	110
The strictfp keyword	111
Comparing objects	111
Three basic coding activities	112
The goto statement	112
Summary	**113**
Certification objectives covered	**114**
Test your knowledge	**114**

Chapter 4: Using Arrays and Collections — 117
Arrays — 118
One-dimensional arrays — 118
- The placement of array brackets — 120
- Initializing arrays — 121
Arrays of objects — 123
Multidimensional arrays — 124
Array techniques — 127
- Traversing arrays — 127
- Comparing arrays — 130
- Copying arrays — 133
- Passing arrays — 137
- Using command-line arguments — 139
The Arrays class — 140
Key points to remember when using arrays — 141
Collections — 142
Iterators — 142
ArrayList — 143
- Creating ArrayList — 144
- Adding elements — 144
- Retrieving elements — 145
- Traversing a ArrayList object — 146
- Sorting a ArrayList object — 148
- Other ArrayList methods — 148
Encapsulating collections — 149
Summary — 150
Certification objectives covered — 151
Test your knowledge — 151

Chapter 5: Looping Constructs — 153
The for statement — 154
The comma operator — 156
The for statement and scope — 157
The for loop variations — 158
The for-each statement — 160
Using the for-each statement with a list — 162
Implementing the Iterator interface — 164
The for-each statement – usage issues — 165
- Null values — 166
- Variable number of arguments — 167
The while statement — 167
The do-while statement — 169
The break statement — 171
The continue statement — 172

Nested loops	**172**
Using labels	**174**
Infinite loops	**175**
Timing is everything	**177**
Pitfalls	**179**
Summary	**182**
Certification objectives covered	**182**
Test your knowledge	**183**
Chapter 6: Classes, Constructors, and Methods	**185**
Classes	**186**
Object creation	186
Memory management	187
Data encapsulation	188
Referencing instance variables	189
Signature	190
Using the this keyword	**190**
Passing parameters	193
Variable number of arguments	196
Immutable objects	198
Constructors	**199**
Default constructors	200
Overloading the constructors	202
Private constructors	204
Constructor issues	205
Java initialization sequence	206
Methods	**207**
Defining methods	207
Calling methods	208
Overloading methods	209
Accessors/mutators	211
Instance and static class members	**212**
Summary	**215**
Certification objectives covered	**215**
Test your knowledge	**216**
Chapter 7: Inheritance and Polymorphism	**219**
Inheritance	**220**
Implementing a subclass	221
Using the protected keyword	223
Overriding methods	225
The @Override annotation	227

Using the final keyword with classes	229
Abstract methods and classes	230
Polymorphism	**231**
Managing classes and objects	**234**
The super keyword and constructors	235
Calling a base class constructor	235
Accessing an overridden method in the base class	237
Determining the type of an object	240
The Object class	241
Casting objects	242
A review of scope	243
Summary	**245**
Certification objectives covered	**246**
Test your knowledge	**246**
Chapter 8: Handling Exceptions in an Application	**249**
Exception types	**251**
Exception handling techniques in Java	**252**
Stack trace	252
Using Throwable methods	253
The traditional try-catch block	**255**
Using the try-with-resource block	**256**
Catch statement	**257**
Order of the catch blocks	258
Using the | operator in a catch block	260
The finally block	**261**
Nested try-catch blocks	**263**
Exception handling guidelines	**264**
Repeating code that threw an exception	264
Not being specific in which exception you are catching	265
Losing the stack trace	265
Scoping and block lengths	267
Throwing a UnsupportedOperationException object	269
Ignoring exceptions	270
Handle exceptions as late as you can	271
Catching too much in a single block	271
Logging exceptions	272
Do not use exceptions to control normal logic flow	273
Do not try to handle unchecked exceptions	274
Summary	**274**
Certification objectives covered	**275**
Test your knowledge	**275**

Chapter 9: The Java Application — 277
Code organization — 277
Packages — 278
The directory/file organization of packages — 278
The import statement — 280
- Avoiding the import statement — 280
- Using the import statement — 280
- Using the wildcard character — 281
- Multiple classes with the same name — 282
- The static import statement — 283
Garbage collection — 283
Resource bundles and the Locale class — 285
Using the Locale class — 286
Using resource bundles — 287
- Using a property resource bundle — 287
- Using the ListResourceBundle class — 290
Using JDBC — 292
Connecting to a database — 292
- Loading a suitable driver — 292
- Establishing a connection — 293
Creating a SQL statement — 293
Handling the results — 294
Summary — 295
Certification objectives covered — 295
Test your knowledge — 296

Appendix: Test Your Knowledge – Answers — 297
Chapter 1: Getting Started with Java — 297
Chapter 2: Java Data Types and Their Usage — 297
Chapter 3: Decision Constructs — 298
Chapter 4: Using Arrays and Collections — 299
Chapter 5: Looping Constructs — 299
Chapter 6: Classes, Constructors, and Methods — 300
Chapter 7: Inheritance and Polymorphism — 300
Chapter 8: Handling Exceptions in an Application — 301
Chapter 9: The Java Application — 301

Index — 303

Preface

You should find this book useful whether you are pursuing Java certification or want to round out your knowledge and gain further confidence in using Java. This book takes a different approach to prepare you for certification. It is designed to provide you with coverage of the topics found in the exam and to provide additional insights in to the use of Java and the development of Java applications. By providing a broader coverage, it goes beyond the immediate certification focus and provides a more comprehensive coverage of the language.

For those pursuing Java certification, the book is organized around the major aspects of Java and addresses the certification topics covered by the Java SE 7 Programmer I (1Z0-803) exam. Each chapter addresses specific certification topics, though some topics are covered in more than one chapter. At the end of each chapter are certification questions that will give you an idea of the nature of the questions you may encounter on the exam. The intent of the book is not to provide an exhaustive set of questions, but rather address those important Java concepts that will prepare you to answer certification questions.

For those of you seeking to advance your knowledge of Java, the book provides insight into Java that you may not have seen before. In particular, the diagrams will hopefully enhance and solidify your understanding of how Java works, especially those figures that describe the use of the program stack and heap. Examples are provided throughout the book that addresses many of the common pitfalls found in developing Java applications.

Regardless of your reasons for reading this book, I hope that you find the book to be rewarding and fulfilling.

What this book covers

Chapter 1, Getting Started with Java, uses an overview of a simple Java application to present the major aspects of Java. The creation of a `customer` class is illustrated, along with the use of getter and setter methods. Also discussed is the development process, the types of Java applications supported, the documentation process in Java, and the use of annotations which have added much to the expressive power of Java.

Chapter 2, Java Data Types and Their Usage, presents the primitive data types available in Java and their corresponding operators. Diagrams are used to explain how the program stack and heap relate to each other and how they affect the scope and lifetime of a variable. In addition, the use of the `String` and `StringBuilder` classes is illustrated and the difference between a class and an object is explained.

Chapter 3, Decision Constructs, focuses on the constructs used to make decisions in Java including the if and switch statements. As these constructs are dependent on logical expression, these types of expressions are covered. The use of the string based switch statement available with Java 7 is illustrated. The correct use of decision constructs is achieved by understanding and avoiding various pitfalls, such as the failure to use block statements and the multitude of problems that can occur when using floating point numbers in comparisons.

Chapter 4, Using Arrays and Collections, focuses on the use of arrays, along with the `Arrays` and `ArrayList` classes. Both single and multidimensional arrays are illustrated. The `Arrays` class is introduced, as it possesses many important methods for manipulating arrays such as filling and sorting arrays. The `ArrayList` class is important as it provides a more flexible container than an array for many problems.

Chapter 5, Looping Constructs, demonstrates the concept of iteration in Java, via constructs such as the while and for loops. These are covered along with common mistakes that can be made when using them. The use of the for-each statement and the iterator is presented, along with coverage of the infinite loop and the break and continue statements.

Chapter 6, Classes, Constructors, and Methods, deals with the creation and use of objects and uses the stack/heap to explain the process. The important Java initialization sequence is discussed. Overloading of constructors and methods are detailed, along with the concepts of signatures, instance/static class members, and immutable objects. Data encapsulation is illustrated throughout the chapter.

Chapter 7, Inheritance and Polymorphism, covers the critical topics of inheritance and polymorphism, with an enhanced discussion of constructors and methods. The use of a signature becomes important again when overriding is used. The power of the `super` keyword is explained in relation to constructors and methods. Scope is re-examined and the concepts of final and abstract classes are explored. The ever present `Object` class is also introduced.

Chapter 8, Handling Exceptions in an Application, covers exception handling including the use of the new try-with-resource block and the | operator in a catch block. Several guidelines and examples dealing with exception handling are provided to help the reader avoid common mistakes in their use.

Chapter 9, The Java Application, examines the use of packages in a Java application. This includes a discussion on the use of the package and import statements including the static import statement. Also discussed is the use of resource bundles to support an application that needs to address the international community and how to connect and use a database using JDBC.

What you need for this book

To work through the examples in this book you will need access to Java 7 SE. This can be downloaded from http://www.oracle.com/technetwork/java/javase/downloads/index.html. The reader may prefer to use an **Integrated Development Environment (IDE)** that supports Java 7 such as NetBeans, Eclipse, or a similar environment.

Who this book is for

This book is for those who are preparing to take the Java SE 7 Programmer I (1Z0-803) exam and/or those who wish to broaden their knowledge about Java.

Conventions

In this book, you will find a number of styles of text that distinguish between different kinds of information. Here are some examples of these styles, and an explanation of their meaning.

Code words in text are shown as follows: "For example, a `person` object and a `square` object can both have a `draw` method."

A block of code is set as follows:

```
public class Application {
   public static void main(String[] args) {
      // Body of method
   }
}
```

Any command-line input or output is written as follows:

```
set path= C:\Program Files\Java\jdk1.7.0_02\bin;%path%
```

 Warnings or important notes appear in a box like this.

 Tips and tricks appear like this.

Reader feedback

Feedback from our readers is always welcome. Let us know what you think about this book—what you liked or may have disliked. Reader feedback is important for us to develop titles that you really get the most out of.

To send us general feedback, simply send an e-mail to feedback@packtpub.com, and mention the book title via the subject of your message.

If there is a topic that you have expertise in and you are interested in either writing or contributing to a book, see our author guide on www.packtpub.com/authors.

Customer support

Now that you are the proud owner of a Packt book, we have a number of things to help you to get the most from your purchase.

Downloading the example code

You can download the example code files for all Packt books you have purchased from your account at http://www.PacktPub.com. If you purchased this book elsewhere, you can visit http://www.PacktPub.com/support and register to have the files e-mailed directly to you.

Errata

Although we have taken every care to ensure the accuracy of our content, mistakes do happen. If you find a mistake in one of our books—maybe a mistake in the text or the code—we would be grateful if you would report this to us. By doing so, you can save other readers from frustration and help us improve subsequent versions of this book. If you find any errata, please report them by visiting http://www.packtpub.com/support, selecting your book, clicking on the **errata submission form** link, and entering the details of your errata. Once your errata are verified, your submission will be accepted and the errata will be uploaded on our website, or added to any list of existing errata, under the Errata section of that title. Any existing errata can be viewed by selecting your title from http://www.packtpub.com/support.

Piracy

Piracy of copyright material on the Internet is an ongoing problem across all media. At Packt, we take the protection of our copyright and licenses very seriously. If you come across any illegal copies of our works, in any form, on the Internet, please provide us with the location address or website name immediately so that we can pursue a remedy.

Please contact us at copyright@packtpub.com with a link to the suspected pirated material.

We appreciate your help in protecting our authors, and our ability to bring you valuable content.

Questions

You can contact us at questions@packtpub.com if you are having a problem with any aspect of the book, and we will do our best to address it.

1
Getting Started with Java

This chapter familiarizes you with basic elements of Java and how to write a simple Java program. A comprehensive understanding of the Java development environment is achieved through simple explanations of the application development process. A Java console program is provided that serves as a starting point and a reference point for this discussion.

In this chapter we will examine:

- What Java is
- The object-oriented development process
- Types of Java applications
- The creation of a simple program
- The definition of classes and interfaces
- Java Application Development
- Java environment
- Java documentation techniques
- The use of annotations in Java
- The core Java packages

Understanding Java as a technology

Sun Microsystems developed the original specifications for the language in the mid 1990s. Patrick Naughton, Mike Sheridan, and James Gosling were the original inventors of Java and the language was called **Oak** at the beginning.

Java is a full-fledged object-oriented programming language. It is platform independent and is normally interpreted rather than compiled like C/C++. It is syntactically and structurally modeled after C/C++ and performs various compile-time and run-time checking operations. Java performs automatic memory management that helps to greatly reduce the problem of memory leaks found in other languages and libraries that dynamically allocate memory.

Java supports many features that, at its time of conception, were not found directly in other languages. These features include threading, networking, security, and **Graphical User Interface (GUI)** development. Other languages could be used to support these capabilities, but they were not integrated in the language to the extent that it was done with Java.

Java uses an independent bytecode that is architecture neutral. That is, it is designed to be machine independent. The byte codes are interpreted and executed by a **Java Virtual Machine (JVM)**. All of its primitive data types are fully specified, as we will see in *Chapter 3, Decision Constructs*. The various releases of the **Java Development Kit (JDK)** and other significant moments are depicted in the following timeline diagram:

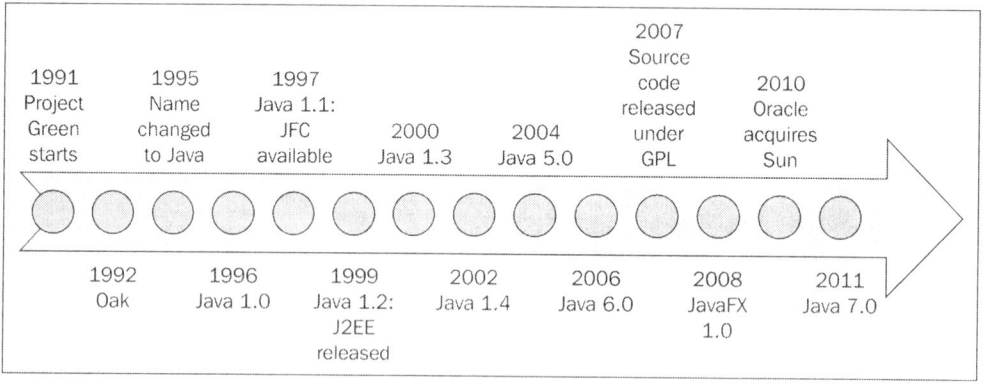

Object-oriented software development

Let's digress for a moment and consider why we are using Java at all. One of the most significant aspects of Java is that it is an **object-oriented (OO)** language. OO technologies are a popular paradigm for developing applications. This approach models an application around a series of real world objects, such as an employee or a ship. In order to solve a problem, it is useful to think of the real world objects that make up the problem domain.

The OO approach is based on three distinct activities:

- **Object Oriented Analysis (OOA)**: This is concerned with determining the functionality of the system, that is, what should the application do
- **Object Oriented Design (OOD)**: This is concerned with how the architecture supports the functionality of the application
- **Object Oriented Programming (OOP)**: This is concerned with the actual implementation of the application

The products of the analysis and design steps are often referred to as analysis and design artifacts. While there may be several different types produced, the one of most interest to the OOP step is called the **class diagram**. The following diagram shows a partial class UML diagram depicting two classes: Customer and CustomerDriver. In the *A simple Java application* section, we will examine the code for these classes. The **Unified Modeling Language (UML)** is a widely used OO technique used to design and document an application. A class diagram is one of the end products of the technique and is used by programmers to create the application:

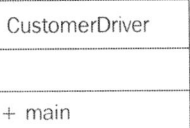

Each box represents a class and is divided into three sections:

- The first section at the top of the box is the name of the class
- The second section lists the variables that make up the class
- The last section lists the class methods

Getting Started with Java

The symbols preceding the variable and method names specify the visibility of these class members. The following are the class diagram symbols used:

- -: Private
- +: Public
- #: Protected (used with inheritance)

Normally, a class diagram consists of many classes and is interconnected with annotated lines showing the relationship between the classes.

The class diagram is intended to clearly show what objects make up the system and how they interact. Once a class diagram is complete it can be implemented using an OOP language such as Java.

> The object-oriented approach is typically used for medium-scale to large-scale projects, where many developers must communicate, and work together, to create an application. For smaller projects involving only a few programmers, such as the one dealt with in most programming classes, the object-oriented approach is not normally used.

OOP principles

While there is some disagreement in what actually makes a programming language an OOP programming language, there are generally three underlying principles that must be supported by an OOP language:

- Data encapsulation
- Inheritance
- Polymorphism

Data encapsulation is concerned with hiding irrelevant information from the users of a class and exposing the relevant. The primary purpose of data encapsulation is to reduce the level of software development complexity. By hiding the details of what is needed to perform an operation, the use of that operation is simpler. How to achieve data encapsulation in Java is explained in the *Access modifiers* section, later in this chapter.

Data encapsulation is also used to protect the internal state of an object. By hiding the variables that represent the state of an object, modifications to the object are controlled through the methods. Any changes to the state are verified by the code in the methods. Also, by hiding variables, sharing of information between classes is eliminated. This reduces the amount of coupling possible in an application.

Inheritance describes the relationship between two classes such that one class re-uses the capabilities of another class. This enables the re-use of software resulting in a more productive developer. Inheritance is covered in detail in *Chapter 7, Inheritance and Polymorphism*.

The third principle is polymorphism and its primary concern is to make the application more maintainable and extendable polymorphism behavior is where the behavior of one or identical methods is dependent upon the object it is executing against. For example, a `person` object and a `square` object can both have a `draw` method. What it draws depends on the object the method is executed against. Polymorphism is discussed in *Chapter 7, Inheritance and Polymorphism*.

These principles are summarized in the following table:

Principle	What it is	Why we use it	How to do it
Data encapsulation	Technique that hides information from the users of that class	To reduce the level of software development complexity	Use access modifiers such as `public`, `private`, and `protected`
Inheritance	Technique to allow a derived or child class to use parts of a base or parent class	To promote the re-use of the software	Use the `extends` keyword
Polymorphism	Technique which supports different behavior of methods that is dependent on the object the method is executing against	To make an application more maintainable	Inherent to the Java language

The `implements` keyword is used in support of polymorphic behavior as is explained in *Chapter 7, Inheritance and Polymorphism*.

Examining the types of Java applications

There are several types of Java applications. These types have allowed Java to flourish in a number of different areas and contributed to Java becoming a very popular programming language. Java is used to develop the following:

- Console and window applications
- Server-based web applications supported by Servlets, JSPs, JSF, and other JEE standards
- Applets that execute within a browser
- Embedded applications
- Componentized building blocks using JavaBeans

While a basic understanding of the types of Java applications is useful in putting Java into context, it also helps to be able to recognize the basic code for these applications. You may not completely understand all of the ins and outs of these application types, but it is useful to see simple code examples.

Reading the code goes a long way towards understanding a language and a specific program. Throughout the book we will use numerous examples to illustrate and explain various aspects of Java. The basic types of Java applications are shown below by presenting short code snippets that are central to that application type.

A simple console application consists of a single class with a `main` method, as shown in the following code snippet:

```java
public class Application {
    public static void main(String[] args) {
        // Body of method
    }
}
```

We will examine this type of application in more depth.

Applets are normally embedded within an HTML page and offer a means of achieving client-side execution of a code. It does not have a `main` method but uses a series of callback methods used by the browser to manage the application. The following code provides an idea of the general appearance of an applet:

```java
import java.applet.*;
import java.awt.Graphics;

public class SimpleApplet extends Applet {

    @Override
```

```
    public void init() {
        // Initialization code
    }

    @Override
    public void paint( Graphics g ) {
        // Display graphics
    }
}
```

The `@Override` annotation is used to ensure that the method that follows is actually overridden. This is discussed in more detail in the *Annotations* section of this chapter.

A **servlet** is a server-side application which renders an HTML page sent to a client. A `doGet` or `doPut` method responds to client-side request. The `out` variable in the following example represents the HTML page. The `println` methods are used to write the HTML code, as shown in the following code snippet:

```
class Application extends HttpServlet {
    public void doGet(HttpServletRequest req,
            HttpServletResponse res)
            throws ServletException, IOException {
        res.setContentType("text/html");

        // then get the writer and write the response data
        PrintWriter out = res.getWriter();
        out.println(
            "<HEAD><TITLE> Simple Servlet</TITLE></HEAD><BODY>");
        out.println("<h1> Hello World! </h1>");
        out.println(
            "<P>This is output is from a Simple Servlet.");
        out.println("</BODY>");
        out.close();
    }
}
```

A **JavaServer Page (JSP)** is actually a disguised Servlet. It provides a more convenient way of developing web pages. The following example uses a JavaBean to display "Hello World" on the web page. The JavaBean is detailed in the following example:

```
<html>
<head>
    <title>A Simple JSP Page</title>
</head>
<body>
```

```
    Hello World!<br/>

    <%
       // This is a scriptlet that can contain Java code
    %>
    <hr>
    <jsp:useBean id="namebean" class="packt.NameBean" scope="session" >
    <jsp:setProperty name="namebean" property="name" value=" Hello world""
    />
    </jsp:useBean>
    <h1> <jsp:getProperty name="namebean" property="name" /></h1>
    </body>
    </html>
```

JavaBeans are building blocks for shared application functionality. They are frequently designed to be used in multiple applications and follow a standard naming convention. The following is a simple JavaBean designed to hold a name (it was used in the previous JSP page):

```
package packt;
public class NameBean {

   private String name= "Default Name"";

   public String getName() {
      return this.name;
   }
   public void setName(String name) {
      this.name = name;
   }
}
```

Enterprise Java Beans (EJB) are components designed to be used in a client/server configuration from a web server. This is a fairly specialized topic that is not relevant to the associate level of certification.

There are several other types of Java technologies such as JSF and Facelets that are a part of JEE. These are improvements over the older Servlet and JSP technologies used to develop web pages.

In this book we will only use simple Java console applications. This type of application is more than sufficient to explain the essence of Java.

Exploring the structure of a Java console program

Let's start with a simple Java program and then use it to explore many of the basic facets of Java. First, a Java application consists of one or more files located somewhere within a filesystem. The name of the files and their locations are both important, as we will see shortly.

> You can download the example code files for all Packt books you have purchased from your account at http://www.PacktPub.com. If you purchased this book elsewhere, you can visit http://www.PacktPub.com/support and register to have the files e-mailed directly to you.

A simple Java application

Our simple program defines a `Customer` class and then uses it in the `CustomerDriver` class as follows:

```java
package com.company.customer;

import java.math.BigDecimal;
import java.util.Locale;

public class Customer {
  private String name;
  private int accountNumber;
  private Locale locale;
  private BigDecimal balance;

  public Customer() {
    this.name = "Default Customer";
    this.accountNumber = 12345;
    this.locale = Locale.ITALY;
    this.balance = new BigDecimal("0");
  }

  public String getName() {
    return name;
  }
  public void setName(String name) throws Exception {
    if(name == null) {
        throw new IllegalArgumentException(
           "Names must not be null");
    } else {
```

```java
      this.name = name;
    }
  }
  public int getAccountNumber() {
    return accountNumber;
  }

  public void setAccountNumber(int accountNumber) {
    this.accountNumber = accountNumber;
  }

  public BigDecimal getBalance() {
    return balance;
  }

  public void setBalance(float balance) {
    this.balance = new BigDecimal(balance);
  }

   public String toString() {
      java.text.NumberFormat format =
         java.text.NumberFormat.getCurrencyInstance(locale);
      StringBuilder value = new StringBuilder();
      value.append(String.format("Name: %s%n", this.name));
      value.append(String.format("Account Number: %d%n",
          this.accountNumber));
      value.append(String.format("Balance: %s%n",
          format.format(this.balance)));
      return value.toString();
   }
}

package com.company.customer;

public class CustomerDriver {

  public static void main(String[] args) {
      // Define a reference and creates a new Customer object
    Customer customer;
    customer = new Customer();
    customer.setBalance(12506.45f);
    System.out.println(customer.toString());
  }
```

The details of how to compile and execute this application are provided in the *Developing Java applications without an IDE* section. When this application is executed you will get the following output:

```
Name: Default Customer
Account number: 12345
Balance: € 12.506,45
```

Understanding the application in detail The following sections detail the significant aspects of the example program. These will be elaborated upon later in more detail in the chapters that follow. Notice, that there are two classes in this application. The `CustomerDriver` class contains the `main` method and is executed first. An instance of the `Customer` class is created and used within the main method.

Package

The package statement specifies the class' `com.company.customer` package. Packages provide a means of grouping similar classes, interfaces, enumerations, and exceptions together. They are discussed in more depth in the *Packages* section in *Chapter 9, The Java Application*.

Import

The `import` statement indicates which packages and classes are used by the class. This allows the compiler to determine whether the package's members are used correctly. Packages need to be imported for all classes, with the exception of the following classes:

- Found in the `java.lang` package
- Located in the current package (`com.company.customer`, in this case)
- Explicitly marked such as `java.text.NumberFormat` as used in the `Customer` class' `toString` method

 The `import` statement informs the compiler of which packages and classes are used by an application and how they can be used.

The Customer class

The first word of the class definition was the keyword, `public`, which is a part of the support Java provides for object-oriented software development. In this context, it specifies that the class is visible outside the package. While not required, it is frequently used for most classes and brings us to the second keyword, `class`, which identifies a Java class.

Instance variables

Four private instance variables were declared next. The use of the `private` keyword hides them from users of the class. The `Locale` class supports applications that can work transparently internationally. `BigDecimal` is the best way of representing currency in Java.

Methods

By making these instance variables private, the designer restricts access to the variables. They are then only accessible through public methods. The combination of private variables and public methods is an example of data encapsulation. If the instance variables are made public instead, other users can directly access the variables. This would improve the efficiency of the program, but may hinder future maintenance efforts. It would be more difficult to change these variables and enforce any sort of validation checks on the changes to the variables.

A series of getter and setter methods were present to return and set the values associated with the private instance variables. This exposes them in a controlled manner. The use of getter and setter methods is a standard approach to achieve encapsulation. For example, trying to assign a null value to a name would throw a `IllegalArmumentException` exception. These types of methods are discussed in the *Method declaration* section.

The `toString` method returns a string representing an instance of a customer. In this case the name, account number, and a localized version of the balance is returned. The use of the `StringBuilder` class is discussed in *Chapter 2, Java Data Types and Their Usage*.

> Methods are found within classes and classes are found within packages.

The CustomerDriver class' main method

The `CustomerDriver` class is referred to as the driver or controller class. Its purpose is to have a `main` method that will create and use other classes.

In a Java application the `main` method is the first method to be executed. If the application consists of multiple classes, normally only one class has a `main` method. A Java application typically needs only one `main` method.

In the `main` method, a new customer is created, a balance is set and then the customer is displayed. A C++ style comment was added to statements to document the declaration and creation of a customer. This was the line beginning with the double forward slashes (`//`). Comments are explained in detail in the *Comments* section.

> The first method that executes in a Java console application is the `main` method.

Exploring the structure of a class

Programming can be thought of as code manipulating data. In Java, code is organized around the following:

- Packages
- Classes
- Methods

Packages are collections of classes with similar functionality. Classes are composed of methods that support the functionality of the class. This organization provides structure to applications. Classes will always be in a package and methods will always be in a class.

> If the package statement is not included in a class definition, the class becomes part of a default package which consists of all of the classes in the same directory that doesn't have a package statement.

Getting Started with Java

Classes, interfaces, and objects

A class is the fundamental building block of object-oriented programs. It generally represents a real-world object. A class definition in Java consists of member variable declarations and method declarations. It begins with the `class` keyword. The body of the class is enclosed with brackets and contains all instance variables and methods:

```
class classname {
  // define class level variables
  // define methods
}
```

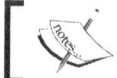 A pair of open and close curly braces constitutes a block statement. This is used in many other parts of Java.

Classes and objects

A class is a pattern or template for creating multiple objects with similar features. It defines the variables and methods of the class. It declares the capabilities of the class. However, before these capabilities can be used, an object must be created. An object is an instantiation of a class. That is, an object is composed of the memory allocated for the member variables of the class. Each object has its own set of member variables.

 The following occurs when a new object is created:

- The new keyword is used to create an instance of a class
- Memory is physically allocated for the new instance of the class
- Any static initializers are executed (as detailed in the *Java initialization sequence* section in *Chapter 6, Classes, Constructors, and Methods*)
- A constructor is called to do initialization
- A reference to the object is returned

The state of an object is typically hidden from the users of the object and is reflected in the value of its instance variables. The behavior of an object is determined by the methods it possesses. This is an example of data encapsulation.

 An object is the instantiation of a class. Each instance of a class has its own unique set of instance variables.

Objects in Java are always allocated on the heap. The heap is an area of memory that is used for dynamically allocated memory, such as objects. In Java, objects are allocated in a program and then released by the JVM. This release of memory is called garbage collection and performed automatically by the JVM. An application has little control over this process. The primary benefit of this technique is the minimization of memory leaks.

A memory leak occurs when memory is dynamically allocated but is never released. This has been a common problem with languages such as C and C++, where it is the responsibility of the programmer to manage the heap.

A memory leak can still occur in Java if an object is allocated but the reference to the object is never released when the object is no longer needed.

Constructors

Constructors are used to initialize an object. Whenever an object is created, a constructor executes. A default constructor is the one that has no arguments and is provided automatically for all classes. This constructor will initialize all instance variables to default values.

However, if the developer provides a constructor, the compiler's default constructor is no longer added. The developer will need to explicitly add a default constructor. It is a good practice to always have a default, no-argument constructor.

Interfaces

An interface is similar to an abstract class. It is declared using the `interface` keyword and consists of only abstract methods and final variables. An abstract class normally has one or more abstract methods. An abstract method is the one that does not have an implementation. It is intended to support polymorphic behavior, as discussed in *Chapter 7, Inheritance and Polymorphism*. The following code defines an interface used to designate a class as capable of being drawn:

```
interface Drawable {
   final int unit = 1;
   public void draw();
}
```

Methods

All executable code executes either within an initializer list or a method. Here, we will examine the definition and use of methods. The initializer list is discussed in *Chapter 6, Classes, Constructors, and Methods*. Methods will always be contained within a class. The visibility of a method is controlled by its access modifiers as detailed in the *Access modifiers* section. Methods may be either static or instance. Here, we will consider instance methods. As we will see in *Chapter 6, Classes, Constructors, and Methods*, static methods typically access static variables that are shared between objects of a class.

Regardless of the type of method, there is only a single copy of a method. That is, while a class may have zero, one, or more methods, each instance of the class (an object) uses the same definition of the method.

Method declaration

A typical method consists of:

- An option modifier
- A return type
- The method name
- A parameter list enclosed in parentheses
- An optional throws clause
- A block statement containing the method's statements

The following `setName` method illustrates these parts of a method:

```java
public void setName(String name) throws Exception {
  if(name == null) {
    throw new Exception("Names must not be null");
  } else {
    this.name = name;
  }
}
```

While the else clause in this example is technically not required, it is a good practice to always use else clauses as it represents a possible execution sequence. In this example, if the if statement's logical expression evaluates to true, then the exception will be thrown and the rest of the method is skipped. Exception handling is covered in detail in *Chapter 8, Handling Exceptions in an Application*.

Methods frequently manipulate instance variables to define the new state of an object. In a well designed class, the instance variables can typically only be changed by the class' methods. They are private to the class. Thus, data encapsulation is achieved.

Methods are normally visible and allow the user of the object to manipulate that object. There are two ways to classify methods:

- **Getter methods**: These methods return the state of an object (also called **accessor methods**)
- **Setter methods**: These are methods that can change the state of an object (also called **mutator methods**)

In the Customer class, setter and getter methods were provided for all of the instance variables, except for the locale variable. We could have easily included a get and set method for this variable but did not, to conserve space.

> A variable that has a get method but not an otherwise visible set method is referred to as a **read-only member variable**. The designer of the class decided to restrict direct access to the variable.
>
> A variable that has a set method but not an otherwise visible get method is referred to as a **write-only member variable**. While you may encounter such a variable, they are rare.

Method signature

The signature of a method consists of:

- The name of the method
- The number of arguments
- The types of the arguments
- The order of the arguments

The signature is an important concept to remember and is used in overloading/overriding methods and constructors as discussed in *Chapter 7, Inheritance and Polymorphism*. A constructor will also have a signature. Notice that the definition of a signature does not include the return type.

The main method

The examples used in the book will be console program applications. These programs typically read from the keyboard and display the output on the console. When a console application is executed by the operating system, the `main` method is executed first. It may then execute other methods.

The `main` method can be used to pass information from the command line. This information is passed to the arguments of the `main` method. It consists of an array of strings representing the program's parameters. We will see this in action in *Chapter 4, Using Arrays and Collections*.

There is only one form of the `main` method in Java, shown as follows:

```
public static void main(String[] args) {
   // Body of method
}
```

The following table shows elements of the main method:

Elements	Meaning
public	The method is visible outside the class.
static	The method can be invoked without creating an object of the class type.
void	The method does not return anything.
args	An array of strings representing the arguments passed.

Returning a value from an application

The `main` method returns `void`, meaning that it is not possible to return a value back to the operating system as part of the normal method invocation sequence. However, it is sometimes useful to return a value to indicate whether the program terminated successfully or not. This is useful when the program is used in a batch type operation where multiple programs are being executed. If one program fails in this execution sequence, then the sequence may be altered. Information can be returned from an application using the `System.exit` method. The following use of the methods will terminate the application and return a zero to the operating system:

```
System.exit(0);
```

> The `exit` method:
> - Forces the termination of all of the application's threads
> - Is extreme and should be avoided
> - Does not provide an opportunity to gracefully terminate the program

Access modifiers

Variables and methods can be declared as one of four types, shown in the following table:

Access type	Keyword	Meaning
Public	`public`	Access is provided to users outside the class.
Private	`private`	Restricts access to members of the class.
Protected	`protected`	Access is provided to classes that inherit the class or are members of the same package.
Package scoped	none	Access is provided to members of the same package.

Most of the time, a member variable is declared as private and a method is declared as public. However, the existence of the other access types implies other potential ways of controlling the visibility of a member. These usages will be examined in *Chapter 7, Inheritance and Polymorphism*.

In the `Customer` class, all of the class variables were declared as private and all of the methods were made public. In the `CustomerDriver` class, we saw the use of the `setBalance` and `toString` methods:

```
customer.setBalance(12506.45f);
System.out.println(customer.toString());
```

As these methods were declared as public, they can be used with the `Customer` object. It is not possible to directly access the balance instance variable. The following statement attempts this:

```
customer.balance = new BigDecimal(12506.45f);
```

The compiler will issue a compile-time error similar to the following:

balance has private access in com.company.customer.Customer

> Access modifiers are used to control the visibility of application elements.

Documentation

The documentation of a program is an important part of the software development process. It explains the code to other developers and provides reminders to the developers of what and why they did what they did.

Documentation is achieved through several techniques. Here, we will address three common techniques:

- **Comment**: This is the documentation embedded in the application
- **Naming conventions**: Following the standard Java naming conventions makes an application more readable
- **Javadoc**: This is a tool used to generate documentation about an application in the form of HTML files

Comments

Comments are used to document a program. They are not executable and are ignored by a compiler. Good comments can go a long way to make a program more readable and maintainable. Comments can be grouped into three types—C style, C++ style, and Java style, as summarized in the following table:

Type of Comment Example	Description
C style	The C style comment uses a two character sequence at the beginning and the end of a comment. This type of comment can extend across multiple lines. The beginning character sequence is a /* while the ending sequence consists of */.

```
/* A multi-line comment
   ...
*/

/* A single line comment */
```

C++ style	The C++ style comment begins with two forward slashes and the comment continues until the end of the line. Essentially everything from the // to the end of the line is treated as a comment.

```
// The entire line is a comment
int total;          // Comment used to clarify variable
area = height*width;   // This computes the area of a rectangle
```

Java Style	The Java style is identical in syntax to the C style comment, except that it starts with /** instead of /*. In addition, special tags can be added within a Java-style comment for documentation purposes. A program called javadoc will read in the source file, which uses these types of comments, and generate a series of HTML files documenting the program. See the *Using Javadocs* section for more detail.

```
/**
 * This method computes the area of a rectangle
 *
 * @param height    The height of the rectangle
 * @param width     The width of the rectangle
 * @return          The method returns the area of a rectangle
 *
 */
public int computeArea(int height, int width)  {
    return height * width;
}
```

Java naming conventions

Java uses a series of naming conventions to make the programs more readable. It is recommended that you always follow this naming convention. By doing so:

- You make your code more readable
- It supports the use of JavaBeans

> More detail on naming conventions can be found at http://www.oracle.com/technetwork/java/codeconvtoc-136057.html.

The Java naming conventions' rules and examples are shown in the following table:

Element	Convention	Example
Package	All letters in lowercase.	`com.company.customer`
Class	First letter of each word is capitalized.	`CustomerDriver`
Interface	First letter of each word is capitalized.	`Drawable`
Variable	First word is not capitalized but the subsequent words are capitalized	`grandTotal`
Method	First word is not capitalized but subsequent words are capitalized. Methods should be verbs.	`computePay`
Constant	Every letter is uppercase.	`LIMIT`

> Following the Java naming conventions is important in maintaining program readability and to support JavaBeans.

Using Javadocs

The Javadoc tool produces a series of HTML files based on the source code and Javadoc tags embedded in the source code. This tool is also distributed with the JDK. While the following example is not an attempt to provide a complete treatment of Javadocs, it should give you a good idea of what Javadocs can do for you:

```
public class SuperMath {
    /**
     * Compute PI - Returns a value for PI.
     *    Able to compute pi to an infinite number of decimal
     *    places in a single machine cycle.
     * @return A double number representing PI
```

```
    */
    public static double computePI() {
        //
    }
}
```

The `javadoc` command, when used with this class, results in the generation of several HTML files. A part of the `index.html` file is shown in the following screenshot:

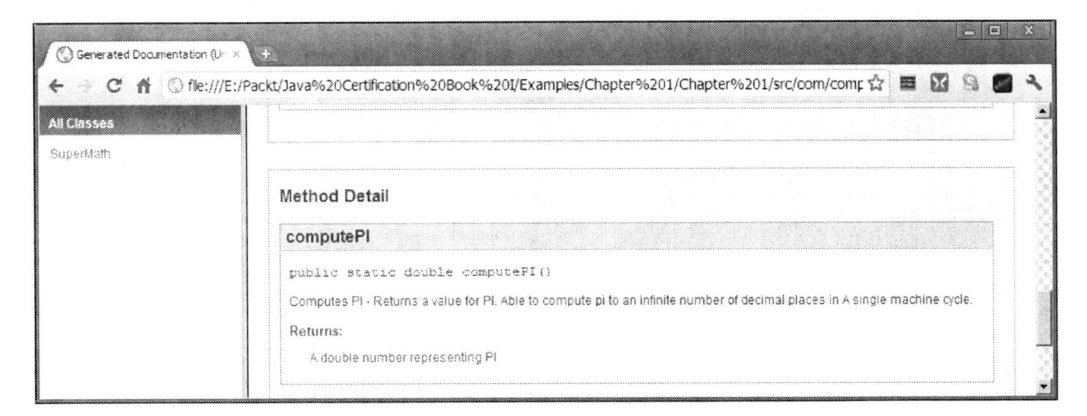

> More information on the use and creation of Javadoc files can be found at http://www.oracle.com/technetwork/java/javase/documentation/index-137868.html.

Investigating the Java application development process

Java source code is compiled to intermediate bytecode. The bytecode is then interpreted at runtime on any platform that has a **Java Virtual Machine (JVM)**. However, this statement is somewhat misleading as Java technology will often compile bytecode directly to machine code. There have been numerous Just-In-Time compiler improvements that speed up the execution of Java applications that often will run nearly as fast as, or sometimes even faster than, natively-compiled C or C++ applications.

Java source code is found in files that end with a `.java` extension. The Java compiler will compile the source code to a bytecode representation and store these bytecodes in a file with a `.class` extension.

Getting Started with Java

There are several **Integrated Development Environments** (IDE) used to support the development of Java applications. A Java application can also be developed from the command line using basic tools from the **Java Development Kit** (JDK).

A production Java application is normally developed on one platform and then deployed to another. The target platform needs to have a **Java Runtime Environment** (JRE) installed on it in order for the Java application to execute. There are several tools that assist in this deployment process. Typically, a Java application is compressed in a **Java Archive** (JAR) file and then deployed. A JAR file is simply a ZIP file with a manifest document embedded within the JAR file. The manifest document often details the content and the type of JAR file being created.

Compiling a Java application

The general steps used to develop a Java application include:

- Create the application using an editor
- Compile it using the Java compiler (javac)
- Execute it using the Java interpreter (java)
- Optionally debug the application as required using a Java debugger

This process is summarized in the following diagram:

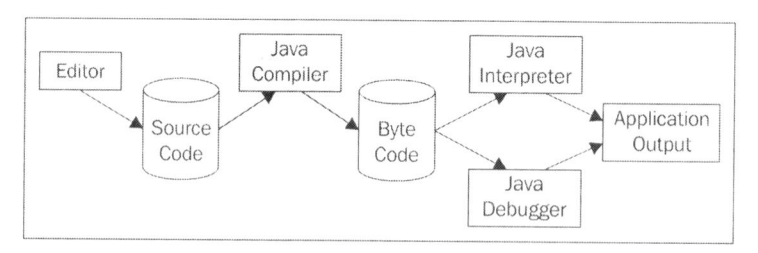

Java source code files are compiled to bytecode files. These bytecode files have a **.class** extension. When a Java package is distributed, the source code files are not normally stored in the same location as the .class files.

SDK file structure

The **Java Software Development Kit (SDK)** can be downloaded and used to create and execute many types of Java applications. The **Java Enterprise Edition (JEE)** is a different SDK and is used to develop enterprise applications typified by web-based applications. The SDK also known as the **Java 2 Enterprise Edition (J2EE)** and you may see it referenced as J2EE. Here, we will only deal with the Java SDK.

While the actual structure of the SDK distribution will vary by release, the typical SDK consists of a series of directories, listed as follows:

- `bin`: This contains the tools used for developing a Java application including the compiler and JVM
- `db`: This is the Apache Derby relational database
- `demo`: This contains a series of demonstration applications
- `include`: This contains header files used to interact with C applications
- `jre`: This is a JRE used by the JDK
- `sample`: This contains sample code for various features of Java

The SDK may include the actual source code for the core classes. This is usually found in the `src.zip` file located under the `JAVA_HOME` root directory.

IDE file structure

Each IDE has a preferred way of organizing the files that make up an application. These organization schemes are not always hard and fast but those presented here are the common ways of arranging the files.

For example, a simple application in the Eclipse IDE consists of two project files and three sub-directories. These files and directories are listed as follows:

- `.classpath`: This is an XML file containing the classpath-related information
- `.project`: This is an XML document describing the project
- `.settings`: This is a directory containing the `org.eclipse.jdt.core.prefs` file which specifies compiler preferences
- `bin`: This is used to contain the package file structure and the application's class files
- `src`: This is used to contain the package file structure and the application's source files

This organization scheme is used by the development tools. The tools often include editors, compilers, linkers, debuggers, and others. These languages frequently use a Make tool to determine which files need to be compiled or otherwise manipulated.

Developing Java applications without an IDE

In this section we will demonstrate how to compile and execute a Java application on a Windows platform using Java 7. The approach is very similar to that used for other operating systems.

Before we can compile and execute the sample program we need to:

- Install the JDK
- Create the appropriate file structure for the application
- Create the files to hold our classes

The latest version of the JDK can be found at `http://www.oracle.com/technetwork/java/javase/downloads/index.html`. Download and install the version that meets your needs. Note the location of the installation, as we will use this information shortly.

As explained earlier, the Java classes must be located in a specific file structure as mirrored in its package name. Create a file structure somewhere in your filesystem that has a top-level directory called `com` under which is a directory called `company` and then under the `company` directory, a directory called `customer`.

In the `customer` directory create two files called `Customer.java` and `CustomerDriver.java`. Use the corresponding classes as found earlier in the *A simple Java application* section.

The JDK tools are found in the JDK directory. When the JDK is installed, the environmental variables are normally set up to allow the successful execution of the JDK tools. However, it is necessary to specify the location of these tools. This is accomplished using the `set` command. In the following command, we set the path environmental variable to reference `C:\Program Files\Java\jdk1.7.0_02\bin` directory, which was the most recent release at the time this chapter was written:

```
set path= C:\Program Files\Java\jdk1.7.0_02\bin;%path%
```

This command prefixes the path to the `bin` directory in front of any previously assigned paths. The `path` environmental variable is used by the operating system to look for the command that is executed at the command prompt. Without this information, the operating system is not aware of the location of the JDK commands.

To compile the program using the JDK, navigate to the directory above the `com` directory. As the classes used as part of this application are part of the `com.company.customer` package we need to:

- Specify the path in the `javac` command
- Execute the command from the directory above the `com` directory

As there are two files that make up this application we need to compile both of them. This can be done using two separate commands as follows:

`javac com.company.customer.Customer.java`

`javac com.company.customer.CustomerDriver.java`

Or, it can be done using a single command and the asterisk wild card character as follows:

`javac com.company.customer.*.java`

The output of the compiler is a bytecode file with the name `CustomerDriver.class`. To execute the program, invoke the Java interpreter with your class file, as shown in the following command. The class extension is not included and its inclusion as part of the filename will result in an error:

`java com.company.customer.CustomerDriver`

The output of your program should be as follows:

```
Name: Default Customer
Account number: 12345
Balance: € 12.506,45
```

Java environment

The Java environment is the operating system and file structure used to develop and execute Java applications. Earlier, we examined the structure of the JDK which are part of the Java environment. Associated with this environment are a series of environmental variables that are used from time-to-time to facilitate various operations. Here, we will examine a few of them in more detail:

- CLASSPATH
- PATH
- JAVA_VERSION
- JAVA_HOME
- OS_NAME
- OS_VERSION
- OS_ARCH

Getting Started with Java

These variables are summarized in the following table:

Name	Purpose	Example
CLASSPATH	Specifies the root directory for classes.	.;C:\Program Files (x86)\Java\jre7\lib\ext\QTJava.zip
PATH	The location of the commands.	
JAVA_VERSION	The version of Java to use.	<param name="java_version" value="1.5.0_11">
JAVA_HOME	The location of the Java directory.	C:\Program Files (x86)\Java\jre6\bin
OS_NAME	The name of the operating system.	Windows 7
OS_VERSION	The version of the operating system	6.1
OS_ARCH	The operating system architecture	AMD64

The CLASSPATH environmental variable is used to identify the root directory of the packages. It is set as follows:

```
c:>set CLASSPATH=d:\development\increment1;%CLASSPATH%
```

The CLASSPATH variable only needs to be set for non-standard packages. The Java compiler will always implicitly append the system's class directories to CLASSPATH. The default CLASSPATH is the current directory and the system's class directories.

There are many other environmental variables associated with an application. The following code sequence can be used to display a list of these variables:

```
java.util.Properties properties = System.getProperties();
properties.list(System.out);
```

A partial display of the output of this code sequence is as follows:

```
-- listing properties --
java.runtime.name=Java(TM) SE Runtime Environment
sun.boot.library.path=C:\Program Files\Java\jre7\bin
java.vm.version=22.0-b10
java.vm.vendor=Oracle Corporation
java.vendor.url=http://java.oracle.com/
path.separator=;
java.vm.name=Java HotSpot(TM) 64-Bit Server VM
...
```

Annotations

Annotations provide information about a program. This information does not reside in the program and does not affect its execution. Annotations are used to support tools such as the compiler and during the execution of the program. For example, the `@Override` annotation informs the compiler that a method is overriding a base class method. If the method does not actually override the base class method because it is misspelled, the compiler will generate an error.

Annotations are applied to elements of the application such as a class, method, or field. It begins with the at sign, `@`, is followed by the name of the annotation, and optionally a list of values enclosed in a set of parentheses.

Common compiler annotations are detailed in the following table:

Annotation	Usage
`@Deprecated`	Used by the compiler to indicate that the element should not be used
`@Override`	The method overrides the base class method
`@SuppressWarnings`	Used to suppress specific compiler warnings

Annotations can be added to an application and used by the third-party tools for specific purposes. It is also possible to write your own annotations when needed.

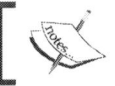

> Annotations are useful in conveying information about an application to tools and the run-time environment

Java class libraries

Java includes a number of libraries of classes that support the development of applications. These include the following, among others:

- `java.lang`
- `java.io`
- `java.net`
- `java.util`
- `java.awt`

These libraries are organized in packages. Each package holds a set of classes. The structure of a package is reflected in its underlying file system. The `CLASSPATH` environmental variable holds the location of packages.

There are a core set of packages that are part of the JDK. These packages provide a crucial element in the success of Java by providing easy access to a standard set of capabilities that were otherwise not readily available with other languages.

The following table shows a list of some of the more commonly used packages:

Package	Usage
java.lang	This is the collection of the basic language types. It includes the root classes, Object and Class, and other items such as threads, exceptions, wrapper, and other fundamental classes.
java.io	Includes streams and random access files.
java.net	Supports sockets, telnet interfaces, and URLs.
java.util	Supports container and utility classes such as Dictionary, HashTable, and Stack. Encoder and decoder techniques such as Date and Time can also be found in this library.
java.awt	Contains the **Abstract Windowing Toolkit** (**AWT**) that contains classes and methods that support a **Graphical User Interface** (**GUI**). It includes classes for events, colors, fonts, and controls.

Summary

In this chapter we examined the essential aspects of Java and a simple Java console application. From a certification standpoint we looked at the structure of a class and a Java application that uses the main method.

We also introduced a number of additional topics that will be covered in more detail in later chapters. This includes the creation of objects and their manipulation, the use of strings and the StringBuilder class, instance and static members of a class, and the use of signatures in the overloading and overriding of a method.

With this foundation we are ready to move on to *Chapter 2, Java Data Types and Their Usage*, where we will examine the nature of variables and how they can be used.

Certification objectives covered

In this chapter we introduced a number of certification topics that will be discussed in more detail in subsequent chapters. Here we covered the following topics in depth:

- Define the structure of a Java class (in the *Exploring the structure of a class* section)
- Create an executable Java application with a main method (in the *Exploring the structure of a Java console program* section)

Test your knowledge

1. What will be printed out if the following code is run with the `java SomeClass hello world` command?

    ```java
    public class SomeClass{
        public static void main(String argv[])
        {
      System.out.println(argv[1]);
        }
    }
    ```

    a. world
    b. hello
    c. hello world
    d. `ArrayIndexOutOfBoundsException` is thrown

2. Consider the following code sequence:

    ```java
    public class SomeClass{
        public int i;
        public static void main(String argv[]){
            SomeClass sc = new SomeClass();
            // Comment line
        }
    }
    ```

    Which of the following statements will compile without syntax or runtime errors if they replace the comment line?

    a. `sc.i = 5;`
    b. `int j = sc.i;`
    c. `sc.i = 5.0;`
    d. `System.out.println(sc.i);`

2
Java Data Types and Their Usage

In this chapter we will learn more about how Java organizes and manipulates data, especially primitive data types and strings. In addition to this, we will explore various related concepts such as scoping and the lifetime of a variable. While strings are not a primitive data type in Java, they are a critical part of many applications and we will examine what Java has to offer.

In this chapter we will focus on:

- The declaration and use of primitive data types
- Using the `String` and `StringBuilder` classes
- How the program stack and heap relate to each other
- The differences between a class and an object
- Constants and literals in Java
- The scope and lifetime of a variable
- Operators, operands, and expressions

Understanding how Java handles data

The core of programming is code that manipulates data. As programmers we are interested in the organization of data and code. The organization of data is referred to as **data structures**. These structures can be static or dynamic in nature. For example, the ages of a population can be stored in consecutive locations in a data structure known as an **array**. While the array data structure has a fixed size, the contents may or may not change. Arrays are discussed in detail in *Chapter 4, Using Arrays and Collections*.

In this section, we will examine several different aspects of variables including:

- How they are declared
- Primitive data types versus objects
- Where they reside in memory
- How they are initialized
- Their scope and lifetime

Java identifiers, objects, and memory

Variables are defined to be of a specific type and are allocated memory. When an object is created, instance variables that make up the object are allocated on the heap. The static variables of an object are allocated to a special area of memory. When a variable is declared as part of a method, the memory for the variable is allocated on the program stack.

Stack and heap

A thorough understanding of the stack/heap and other issues is critical for understanding how a program works, and in turn how well a developer can use a language such as Java to do his job. These concepts provide a framework for understanding how an application works and are the basis for the implementation of the runtime system used by Java, not to mention almost every other programming language in existence.

With this said, the concept of a stack and heap is fairly simple. The **stack** is an area where the parameters of a method and its local variables are stored each time a method is invoked. The **heap** is an area of memory where objects are allocated when the new keyword is invoked. The parameters and local variables of a method make up an **activation record**, also called a **stack frame**. Activation records are pushed onto a stack when the method is invoked and popped off the stack when the method returns. The temporary existence of these variables determines the lifetime of the variables.

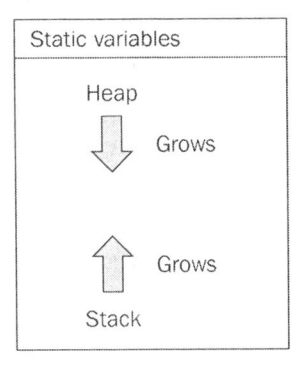

```
Static variables

       Heap

        ⬇  Grows


        ⬆  Grows
       Stack
```

The stack grows toward the heap when a method is invoked and shrinks when the method returns. The heap does not grow in a predictable order and can become fragmented. As they share the same memory space, if the heap and stack collide then the program will terminate.

> Understanding the concept of stack and heap is important because:
> - It provides a foundation used to understand how data is organized in an application
> - It helps explain the concept of the scope and lifetime of a variable
> - It helps explain how recursion works

We will re-use the program illustrated in *Chapter 1, Getting Started with Java*, to demonstrate the use of stack and heap. The program has been duplicated here for your convenience:

```java
package com.company.customer;
import java.math.BigDecimal;
import java.util.Locale;

public class Customer {
    private String name;
    private int accountNumber;
```

```java
    private Locale locale;
    private BigDecimal balance;

    public Customer() {
      this.name = "Default Customer";
      this.accountNumber = 12345;
      this.locale = Locale.ITALY;
      this.balance = new BigDecimal("0");
    }

    public String getName() {
      return name;
    }
    public void setName(String name) throws Exception {
      if(name == null) {
        throw new Exception("Names must not be null");
      } else {
        this.name = name;
      }
    }

    public int getAccountNumber() {
      return accountNumber;
    }

    public void setAccountNumber(int accountNumber) {
      this.accountNumber = accountNumber;
    }

    public BigDecimal getBalance() {
      return balance;
    }

    public void setBalance(float balance) {
      this.balance = new BigDecimal(balance);
    }

    public String toString() {
      java.text.NumberFormat format;
      format = java.text.NumberFormat.getCurrencyInstance(locale);
      return format.format(balance);
    }
  }
```

```
package com.company.customer;
public class CustomerDriver {
  public static void main(String[] args) {
    Customer customer;       // defines a reference to a Customer
    customer = new Customer();  // Creates a new Customer object
    customer.setBalance(12506.45f);
    System.out.println(customer.toString());
  }
}
```

When the `main` method is executed, an activation record is pushed onto the program stack. As shown in the following diagram, its activation record consists of only the single `args` parameter and the `customer` reference variable. When the instance of the `Customer` class is created, an object is created and allocated on the heap. The state of the stack and heap reflected in this example occurs after the `Customer` constructor executes. The `args` reference variable points to an array. Each element of the array references a string representing the applications' command line arguments. In the example shown in the following diagram, we assume there are two command line arguments, Argument 1 and Argument 2:

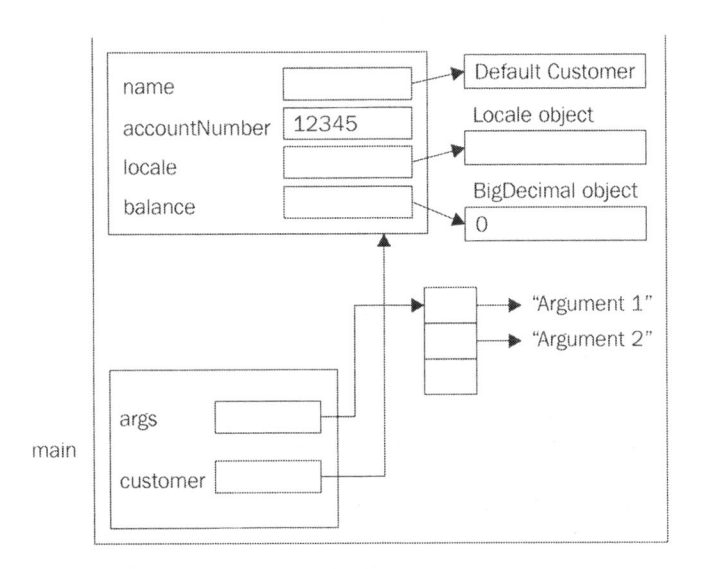

When the `setBalance` method is executed, its activation record is pushed onto the program stack as illustrated below. The `setBalance` method has a single parameter, `balance`, which is assigned to the `balance` instance variable. But first, it is used as an argument to the `BigDecimal` constructor. The `this` keyword references the current object.

Java Data Types and Their Usage

Heap is the memory that is dynamically allocated for objects. A heap manager controls how this memory is organized. When an object is no longer needed, a garbage collection routine will execute to free up the memory so it can be re-used. Before an object is disposed of, the object's `finalize` method is executed. However, there is no guarantee that the method will execute as the program may terminate without the need for the garbage collection routine to run. The original `BigDecimal` object will be eventually destroyed.

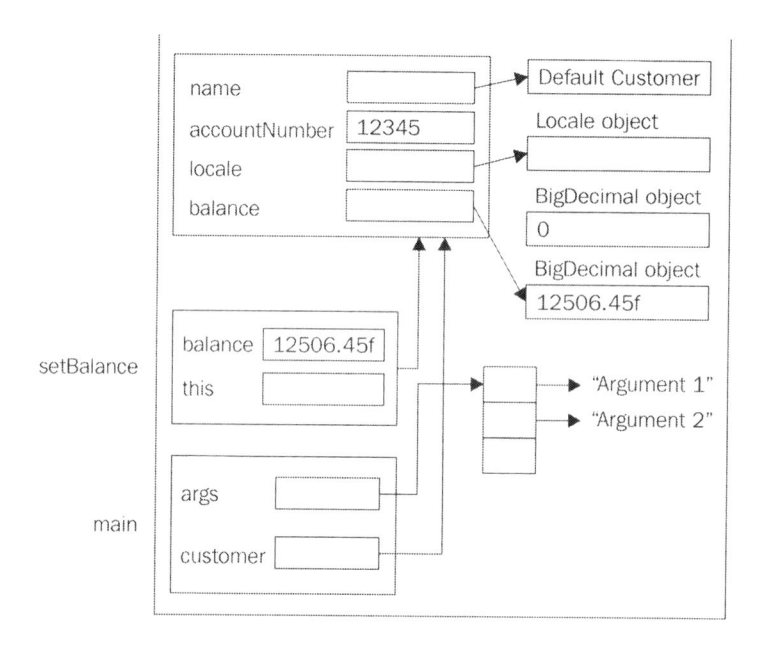

> In C++, when an object is about to be destroyed its destructor will be executed. The closest thing Java has to this is the `finalize` method, which will execute when the object is processed by the garbage collector. However, the garbage collector may not run and, thus, the `finalize` method may never execute. This paradigm shift results in an important difference in how we manage resources. The try-with-resources block introduced in *Chapter 8, Handling Exceptions in an Application*, offers a technique for dealing with this situation.

Declaring a variable

A variable is also called an identifier. The term "variable" implies that its value can be changed. This is usually the case. However, if the identifier is declared as a constant, as discussed in the *Constants* section, then it is not really a variable. Regardless of this, the terms variable and identifier are normally considered to be synonymous.

The declaration of a variable begins with the data type and is followed by the variable name and then a semicolon. The data type may be a primitive data type or a class. When the data type is a class, the variable is an object reference variable. That is, it is a reference to an object.

> A reference variable is effectively a disguised C pointer.

Variables can be classified into the following three categories:

- Instance variables
- Static variables
- Local variables

Instance variables are used to reflect the state of the object. Static variables are variables that are common to all instances. Local variables are declared within a method and are visible only within the block in which they are declared.

Identifiers are case-sensitive and can only be composed of:

- Letters, numbers, the underscore (_) and the dollar sign ($)
- Identifiers may only begin with a letter, the underscore or a dollar sign

Examples of valid variable names include:

- `numberWheels`
- `ownerName`
- `mileage`
- `_byline`
- `numberCylinders`
- `$newValue`
- `_engineOn`

By convention, identifiers and methods start with the first word in lower case and subsequent words capitalized as discussed in the *Java naming conventions* section of *Chapter 1, Getting Started with Java*. Examples of conventional declarations include the following:

- `int numberWheels;`
- `int numberCylinders;`
- `float mileage;`
- `boolean engineOn;`
- `int $newValue;`
- `String ownerName;`
- `String _byline;`

In the preceding examples, each variable, except the last two, is declared as a primitive data type. The last one is declared as a reference to a `String` object. The reference variable can reference a `String` object but, in this example, is assigned a `null` value meaning that it is does not currently reference a string. Strings are covered in more detail in the *The String class* section. The following code snippet declares three variables of type integer:

```
int i;
int j;
int k;
```

It is also possible to declare all three variables on a single line, shown as follows:

```
int i, j, k;
```

Primitive data types

There are eight primitive data types defined in Java, as listed in the following table. In Java, the size of each of the data types is the same for all machines:

Data type	Size in bytes	Internal representation	Range
boolean	--	Not precisely defined	`true` or `false`
byte	1	8-bit two's complement	-128 to +127
char	2	Unicode	`\u0000` to `\uffff`
short	2	16-bit two's complement	-32768 to 32767
int	4	32-bit two's complement	-2,147,483,648 to 2,147,483,647

Data type	Size in bytes	Internal representation	Range
long	8	64-bit two's complement	-9,223,372,036,854,775,808 to 9,223,372,036,854,775,807
float	4	32-bit IEEE 754 floating point	3.4e +/- 38 (7 digits)
double	8	64-bit IEEE 754 floating point	1.7e +/- 308 (15 digits)

The `String` data type is also a part of Java. While it is not a primitive data type, it is a class and is discussed in detail in the *The String class* section.

Another common data type is currency. There are several ways of representing money in Java as detailed in the following table. However, the recommended approach is to use the `BigDecimal` class.

Data type	Advantages	Disadvantages
Integer	Good for simple currency units, such as a penny.	It does not use a decimal point, such as that used in dollars and cents.
Floating point	It uses a decimal point.	Rounding errors are very common.
`BigDecimal` class	• Handles large numbers. • Uses decimal points. • Has built-in rounding modes.	More difficult to use.

When using `BigDecimal`, it is important to note the following:

- Use the constructor with the `String` argument as it does a better job at placing the decimal point
- `BigDecimal` is immutable
- The `ROUND_HALF_EVEN` rounding mode introduces the least bias

The `Currency` class is used to control the formatting of currency.

> Another recommendation regarding currency representation is based on the number of digits used.
>
> | Number of digits | Recommended data type |
> |---|---|
> | Less than 10 | Integer or `BigDecimal` |
> | Less than 19 | Long or `BigDecimal` |
> | Greater than 19 | `BigDecimal` |

Floating point numbers can be a significant source of problems in most languages. Consider the following snippets where we add `0.1` in an attempt to get the value `1.0`:

```
float f = 0.1f;
for(int i = 0; i<9; i++) {
    f += 0.1f;
}
System.out.println(f);
```

The output is as follows:

`1.0000001`

It reflects the fact that the decimal value `0.1` cannot be accurately represented in base two. This means we must always be vigilant when working with floating point numbers.

Wrapper classes and autoboxing

Wrapper classes are used to enclose primitive data type values inside an object. Prior to the availability of boxing, there were often situations where it was necessary to explicitly use wrapper classes such as the `Integer` and `Float` classes. This was required to be able to add primitive data types to collections often found in the `java.util` package, including the `ArrayList` class, because methods of these data classes used objects as arguments. Wrapper classes include the following data types:

- Boolean
- Byte
- Character
- Short
- Integer
- Long
- Float
- Double

Objects of these wrapper classes are immutable. That is, their values cannot be changed.

Autoboxing is the automatic conversion of primitive data types into their corresponding wrapper classes. This is performed as needed so as to eliminate the need to perform trivial, explicit conversion between primitive data types and their corresponding wrapper classes. **Unboxing** refers to the automatic conversion of a wrapper object to its equivalent primitive data type. In effect, primitive data types are treated as if they are objects in most situations.

There are a few things to remember when working with primitives and objects. First, objects can be `null`, whereas primitives cannot be assigned a `null` value. This can present problems from time to time. For example, unboxing a null object will result in a `NullPointerException`. Also, be careful when comparing primitives and objects when boxing does not occur as illustrated in the following table:

Comparison	Two primitives	Two objects	One of each
`a == b`	Simple comparison	Compares reference values	Treated as two primitives
`a.equals(b)`	Will not compile	Compares for equality of values	Will not compile if a is a primitive, otherwise their values are compared

Initializing identifiers

The initialization of Java variables is actually a complex process. Java supports four ways of initializing variables:

- Default initial values
- Instance variable initializers
- Instance initializers
- Constructors

In this chapter we will examine the first two approaches. The latter two techniques are covered in *Chapter 6, Classes, Constructors, and Methods*, where the whole initialization process is put together.

Java Data Types and Their Usage

When explicit values are not provided, initial default values are used as the object is created. In general, when a field of an object is allocated it is initialized to a zero value as detailed in the following table:

Data type	Default value (for fields)
boolean	false
byte	0
char	'\u0000'
short	0
int	0
long	0L
float	0.0f
double	0.0d
String (or any object)	null

For example, in the following class, name is assigned null and age has a value of 0:

```
class Person {
  String name;
  int age;
  ...
}
```

The instance variable initializers' operator can be used to explicitly assign a value to a variable. Consider the following variation of the Person class:

```
class Person {
  String name = "John Doe";
  int age = 23;
  ...
}
```

When an object of type Person is created, the name and age fields are assigned the values John Doe and 23 respectively.

However, when a local variable is declared, it is not initialized. It is, therefore, important to either use the initialization operator when declaring the variable or to not use the variable until a value has been assigned to it. Otherwise, a syntax error will result.

Java constants, literals, and enumerations

Constants and literals are similar in that they cannot be changed. A variable can be declared using the `final` keyword as a primitive data type that cannot change and is, thus, referred to as a constant. A literal is a token such as `35` or `'C'` which represents a value. Obviously, it cannot be modified either. Related to this concept are immutable objects—objects which cannot be modified. While the object cannot be modified, the reference variable pointing to the object can be changed.

Enumerations are also effectively constant in nature. They are used to provide a convenient way of dealing with sets of values as a list. For example, an enumeration can be created to represent the suits of a card deck.

Literals

Literal constants are simple numbers, characters, and strings that represent a quantity. There are three basic types:

- Numeric
- Character
- Strings

Numeric literals

Numeric constants consist of a series of digits with an optional sign and an optional decimal point. Numeric literals that contain a decimal point are by default `double` constants. Numeric constants can also be prefixed with a `0x` to indicate the number is a hexadecimal number (base 16). Numbers that begin with a `0` are octal numbers (base 8). A suffix of `f` or `F` can be used to declare a floating point literal as of type `float`.

Numeric literal	Base	Data type	Decimal equivalent
25	10	int	25
-235	10	int	-235
073	8	int	59
0x3F	16	int	63
23.5	10	double	23.5
23.5f	10	float	23.5
23.5F	10	float	23.5
35.05E13	10	double	350500000000.00

Integer literals are common. Normally they are expressed in base 10, but octal and hexadecimal literals can be created using the appropriate prefix. Integer literals are of type `int` by default. A literal can be specified as type `long` by appending an L to the end of the literal. The following table illustrates literals and their corresponding data types:

Literal	Type
45	int
012	An integer expressed as an octal number.
0x2FFC	An integer expressed as a hexadecimal number.
10L	long
0x10L	A long expressed as a hexadecimal number.

> Either a lowercase or uppercase L can be used to designate an integer as type long. However, it is better to use an uppercase L to avoid confusing the letter with the numeric digit "1" (one). In the following example, an unwary reader might see the literal as one hundred and one versus the integer 10:
>
> 101 versus 10L

Floating point literals are numbers that contain a decimal point or those that are written using scientific notation.

Literal	Type
3.14	double
10e6	double
0.042F	float

Java 7 added the ability to uses underscore characters (_) in numeric literals. This enhances the readability of code by adding a visual spacing between significant parts of a literal. Underscores can be added almost anywhere with a numeric literal. It can be used with floating point numbers and with any integer base (binary, octal, hexadecimal, or decimal). In addition, base 2 literal are also supported.

The following table illustrates the use of underscores with a variety of numeric literals:

Example	Usage
111_22_3333	Social security number
1234_5678_9012_3456	Credit card number
0b0110_00_1	Binary literal representing a byte
3._14_15F	PI
0xE_44C5_BC_5	Hexadecimal literal for a 32-bit quantity
0450_123_12	24-bit octal literal

The use of literals in code has no effect on the internal representation of the number or how it is displayed. For example, if we use a long literal to represent a social security number, the number is stored internally in two's complement notation and will be displayed as an integer:

```
long ssn = 111_22_3333L;
System.out.println(ssn);
```

The output is as follows:

```
111223333
```

If it is necessary to display the number formatted as a social security number, this needs to be done in code. The following is one of the approaches to this:

```
long ssn = 111_22_3333L;
String formattedSsn = Long.toString(ssn);
for (int i = 0; i < formattedSsn.length(); i++) {
    System.out.print(formattedSsn.charAt(i));
    if (i == 2 || i == 4) {
        System.out.print('-');
    }
}
System.out.println();
```

When executed, we get the following output:

```
111-22-3333
```

The use of the underscore is to make the code more readable to the developer but it is ignored by the compiler.

Java Data Types and Their Usage

There are a couple of other things to consider when using underscores in literals. First, consecutive underscores are treated as one and also ignored by the compiler. Also, underscores cannot be placed:

- At the beginning or end of a number
- Adjacent to a decimal point
- Prior to the D, F, or L suffix

The following table illustrates invalid usage of the underscores. These will generate the syntax error: `illegal underscore`:

Example	Problem
_123_6776_54321L	Cannot begin with an underscore
0b0011_1100_	Cannot end with an underscore
3._14_15F	Cannot be adjacent to a decimal point
987_654_321_L	Cannot be adjacent to an L suffix

Some applications need to manipulate the bits of a value. The following example will perform a bitwise AND operation against a value using a mask. A mask is a sequence of bits that are used to isolate part of another value. In this example, `value` represents a bit sequence whose last four bits we wish to isolate. The binary literal represents the mask:

```
value & 0b0000_11111;
```

The AND operation will return zeroes when ANDed with a mask containing zeroes. In the preceding example, the first four bits of result of the expression will be zeroes. The last four bits are ANDed with ones which result in the last four bits of the result being the same as the last four bits of value. Thus, the last four bits have been isolated.

This is illustrated by executing the following code sequence:

```
byte value = (byte) 0b0111_1010;
byte result = (byte) (value & 0b0000_1111);
System.out.println("result: " +
   Integer.toBinaryString(result));
```

When executed we get the following output:

`result: 1010`

The following diagram illustrates this AND operation:

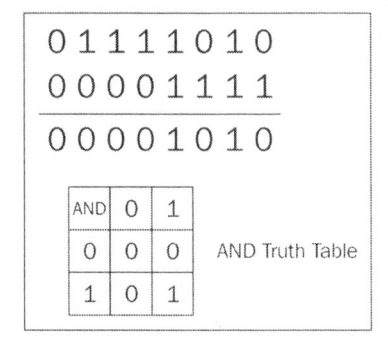

```
0 1 1 1 1 0 1 0
0 0 0 0 1 1 1 1
─────────────────
0 0 0 0 1 0 1 0
```

AND	0	1
0	0	0
1	0	1

AND Truth Table

Character literals

Character literals are single characters enclosed in single quotes.

```
char letter = 'a';
letter = 'F';
```

However, one or more symbols can be used to represent a character. The backslash character is used to "escape" or give special meaning to a letter. For example, '\n' represents the carriage return line feed character. These special escape sequences represent certain special values. These escape sequences can also be used within a string literal. Escape sequence characters are listed in the following table:

Escape Sequence Character	Meaning
\a	alert (bell)
\b	backspace
\f	form feed
\n	new line
\r	carriage return
\t	horizontal tab
\v	vertical tab
\\	backslash
\?	question mark
\'	single quote
\"	double quote
\ooo	octal number
\xhh	hexadecimal number

String literals

String literals are a sequence of characters that are enclosed in a set of double quotes. String literals cannot be split across two lines:

```
String errorMessage = "Error - bad input file name";
String columnHeader = "\tColumn 1\tColumn2\n";
```

Constants

Constants are identifiers whose values cannot change. They are used in situations where, instead of using a literal, a more readable name should be used instead. In Java, constants are declared by prefixing the variable declaration with the `final` keyword.

In the following example, three constants are declared—PI, NUMSHIPS, and RATEOFRETURN. Each of these is capitalized as per the standard *Java naming conventions* section of *Chapter 1, Getting Started with Java*, and is given a value. These values cannot be changed:

```
final double PI = 3.14159;
final int NUMSHIPS = 120;
final float RATEOFRETURN = 0.125F;
```

In the following statement, an attempt is made to change the value of PI:

```
PI = 3.14;
```

Depending on the compiler, an error message similar to the following will be generated:

`cannot assign a value to final variable PI`

This means you cannot change the value of the constant variable.

> Constants provide other benefits apart from always having the same value. A constant number or object can be handled more efficiently and optimized. This makes the application that uses them more efficient and easier to understand. Instead of using 3.14159 every place it is needed, we can simply use PI.

The final keyword

While the `final` keyword is used to declare a constant, it has other uses as detailed in the following table. We will cover its use with methods and classes in later chapters:

Applied to	Meaning
Primitive data declaration	The value assigned to the variable cannot be changed.
Reference variable	You cannot change the variable to reference a different variable. However, it may be possible to change the object the variable is referencing to.
Method	The method cannot be overridden.
Class	The class cannot be extended.

Enumerations

Enumerations are actually subclasses of the `java.lang.Enum` class. In this section, we will look at the creation of a simple enumeration. A more complete treatment of this topic is found in *Chapter 6, Classes, Constructors, and Methods*.

The following example declares an enumeration called `Directions`. This enumeration represents the four cardinal points.

```
public enum Directions {NORTH, SOUTH, EAST, WEST}
```

We can declare a variable of this type and then assign values to it. The following code sequence illustrates this:

```
Directions direction;
direction = Directions.EAST;
System.out.println(direction);
```

The output of this sequence is as follows:

EAST

A `enum` call also can be used as part of a switch statement illustrated as follows:

```
switch(direction) {
case NORTH:
  System.out.println("Going North");
  break;
case SOUTH:
  System.out.println("Going South");
  break;
```

```
    case EAST:
      System.out.println("Going East");
      break;
    case WEST:
      System.out.println("Going West");
      break;
  }
```

When executed with the previous code we get the following output:

`Going East`

Immutable objects

Immutable objects are objects whose fields cannot be modified. There are several classes whose objects are immutable in the Java core SDK including the `String` class. Perhaps surprisingly, the `final` keyword is not used for this purpose. These are discussed in more detail in *Chapter 6, Classes, Constructors, and Methods*.

Instance versus static data

There are two different types of variables (data) within a class: instance and static. When an object is instantiated (using the `new` keyword with a class name), each object is composed of the instance variables that make up that class. However, there is only one copy of static variables allocated for each class. While each class has its own copy of instance variables, all of the classes share a single copy of the static variables. These static variables are allocated to a separate area of memory and exist for the lifetime of the class.

Consider the addition of a common discount percentage that may be applied selectively to some, but not all, customers. Regardless of whether or not it is applied, the percentage is always the same. Based on these assumptions, we can add a static variable to a class as follows:

```
    private static float discountPercentage;
```

Static methods and fields are covered in more detail in *Chapter 6, Classes, Constructors, and Methods*.

Scope and lifetime

Scope refers to where in a program a specific variable can be used. In general, a variable is visible within the block statement in which it is declared but not outside it. A block statement is a sequence of code encapsulated by curly braces.

If a variable is in scope, then it is visible to the code and can be accessed. If it is not in scope, then the variable cannot be accessed and any attempts to do so will result in a compile-time error.

The lifetime of a variable refers to the time period in which it has been allocated memory. When a variable is declared as a local variable of a method, the memory allocated to the variable is in the activation record. As long as the method has not returned, the activation record exists and memory is allocated for the variable. As soon as the method returns, the activation record is removed from the stack and the variable is no longer in existence and cannot be used.

The lifetime of an object allocated from the heap begins when the memory is allocated and ends when the memory is de-allocated. In Java, memory is allocated for an object using the `new` keyword. An object and its memory are marked for de-allocation when it is no longer referenced. In reality, it is de-allocated at some indeterminate point in the future when a garbage collection routine runs, if at all. If an object has no references to it, it can be used or accessed even if the garbage collector has not reclaimed it.

Scoping rules

Scoping rules are critical to understanding how block structure languages, such as Java, work. These rules explain when a variable can be used and which one of the several will be used when a naming conflict occurs.

Scoping rules revolve around the concept of a block. A block is delineated by an opening and closing curly brace. These blocks are used to group code together and to define the scope of a variable. The following diagram shows the scope of three variables, i, j, and k:

```
private void demo ( ) {
    int i;
    i = 100;

    for (int j=0; j<100; j++) {
        a[ j ] = 0;                    Scope of j
        b[ j ] = -1;
    }                                                    Scope of i

    while (i>0) {
        int tmp;
        tmp = i * i;                   Scope of tmp
        a[i] = b[i] * i + tmp;
    }
}
```

Access modifiers

Access modifiers can be used as prefixes when declaring instance and static variables and methods. Modifiers are applied in various combinations to provide specific behaviors. The modifier order is not always important, but consistent style leads to more readable code. All modifiers are optional though there are some default modifiers. Access modifiers include:

- `public`: A public object is visible to all methods inside and outside its own class.
- `protected`: This allows protection between the current class and subclasses. A protected object is invisible outside *-the class, fully visible to subclasses.
- `private`: A private variable cannot be seen by any class other than the one in which it is defined (including subclasses).
- **package**: This visibility is the default protection. Only classes within the package have the access (public within the package).

To explain the scope of variables, consider the package/class organization shown in the following diagram, where the arrows indicate inheritance:

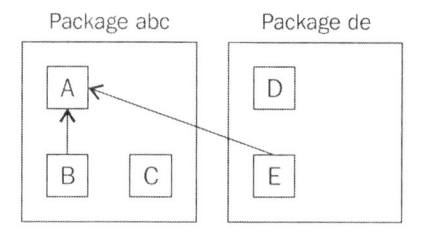

Assume that the A class is defined as follows:

```
public class A{
   public int  publicInt;
   private int privateInt;
   protected int  protectedInt;
   int defaultInt;  // default (package)
}
```

All of the variables are of type `int`. The `publicInt` variable is a public variable. It can be seen by all methods inside and outside of this class. The `privateInt` variable is only visible within this class. The `protectedInt` variable is visible only to those classes within this package. The `protectedInt` variable is visible to this class, its subclasses, and other classes in the same package. It is not visible elsewhere. The following table shows the visibility of each of the declaration types to each of the classes:

	A	B	C	D	E
`publicInt`	Visible	Visible	Visible	Visible	Visible
`privateInt`	Visible	Invisible	Invisible	Invisible	Invisible
`protectedInt`	Visible	Visible	Visible	Invisible	Visible
`defaultInt`	Visible	Visible	Visible	Invisible	Invisible

Data summary

The following table summarizes the types of variables and their relationship to Java compile-time and run-time elements:

Program element	Variable type	Part of	Allocated to
Class	Instance	Object	Heap
	Static	Class	Special region of memory
Method	Parameter	Activation record	Activation record of the stack
	Local		

Building expressions using operands and operators

An expression consists of operands and operators. Operands are normally variable names or literals while operators act on operands. The following are examples of expressions:

```
int numberWheels = 4;
System.out.println("Hello");
numberWheels = numberWheels + 1;
```

Java Data Types and Their Usage

There are several ways of classifying operators:

- Arithmetic
- Assignment
- Relational
- Logical complement
- Logical
- Conditional
- Bitwise

Expressions can be thought of as the building blocks of a program. They are used to express the logic of the program.

Precedence and associativity

Java operators are summarized in the following precedence and associativity table. Most of these operators are straightforward:

Precedence	Operator	Associativity	Operator
1	++	Right	Pre/post increment
	--	Right	Pre/post decrement
	+, -	Right	Unary plus or minus
	~	Right	Bitwise complement
	!	Right	Logical complement
	(cast)	Right	Cast
2	*, /, and %	Left	Multiplication, division, and modulus
3	+ and -	Left	Addition and subtraction
	+	Left	String concatenation
4	<<	Left	Left shift
	>>	Left	Right shift and sign fill
	>>>	Left	Right shift and zero fill
5	<, <=, >, >=	Left	Logical
	Instanceof	Left	Type comparison
6	== and !=	Left	Equality and inequaltity
7	&	Left	Bitwise and Boolean AND
8	^	Left	Bitwise and Boolean XOR
9	|	Left	Bitwise and Boolean OR

Precedence	Operator	Associativity	Operator
10	&&	Left	Boolean AND
11	||	Left	Boolean OR
12	?:	Right	Conditional
13	=	Right	Assignment
	+=, -=, *=, /=, and %=	Right	Compound

While the use of most of these operators is straightforward, more detailed examples for their usage are provided in later chapters. However, bear in mind that there are no other variations and no other operators available in Java. For example, += is a valid operator while =+ is not. However, it can be used with possibly unintended consequences. Consider the following:

```
total = 0;
total += 2;  // Increments total by 2
total =+ 2;  // Valid but simply assigns a 2 to total!
```

The last statement appears to be using a =+ operator. In reality, it is the assignment operator followed by the unary plus operator. A +2 is assigned to total. Remember, Java ignores white space except within string literals.

Casting

When one type of data is assigned to a different type of data, it is possible to lose information. If the data is being assigned from a more precise data type to a less precise data type, it is referred to as **narrowing**. For example, if the floating point number 45.607 is assigned to an integer, the fractional part, .607, is lost.

When making assignments of this type, the cast operator should be used. The cast operator is simply the data type that you are casting to, enclosed in parentheses. The following shows several explicit casting operations:

```
int i;
float f = 1.0F;
double d = 2.0;

i = (int) f;    // Cast a float to an int
i = (int) d;    // Cast a double to an int
f = (float) d;  // Cast a double to a float
```

Without the use of the cast operator in such situations, the compiler will issue a warning. The warning is there to suggest that you look more closely at the assignments. The loss of precision may or may not be a problem, depending upon the use of the data within the application. Without a cast operator, an implicit cast is made when the code is executed.

Working with characters and strings

The primary classes include the `String`, `StringBuffer`, `StringBuilder`, and `Character` classes. There are several other classes and interfaces related to string and character manipulation, listed as follows, that you should be aware of. However, not all of the following classes will be detailed here.

- `Character`: This deals with the manipulation of character data
- `Charset`: This defines a mapping between Unicode characters and a sequence of bytes
- `CharSequence`: In this, an interface is implemented by the `String`, `StringBuffer` and `StringBuilder` classes defining common methods
- `StringTokenizer`: This is used for tokenizing text
- `StreamTokenizer`: This is used for tokenizing text
- `Collator`: This is used to support operations on locale specific strings

The String, StringBuffer, and StringBuilder classes

There are several string-related classes available to the Java programmer. In this section, we will examine the classes and techniques available in Java for manipulating this type of data.

The three primary classes for string manipulation found in the JDK are the `String`, `StringBuffer`, and `StringBuilder`. The `String` class is the most widely used of these classes. The `StringBuffer` and `StringBuilder` classes were introduced in Java 5 to address efficiency issues of the `String` class. The `String` class is immutable and an application that requires frequent changes to a string will be burdened by the overhead of having to create new immutable objects. The `StringBuffer` and `StringBuilder` classes are mutable objects and can be used more efficiently when strings need to be modified frequently. `StringBuffer` differs from `StringBuilder` in that its methods are synchronized.

In terms of methods supported by the classes, the methods of `StringBuffer` and `StringBuilder` are identical. They only differ in whether the methods are synchronized or not.

Class	Mutable	Synchronized
`String`	No	No
`StringBuffer`	Yes	Yes
`StringBuilder`	Yes	No

A synchronized method is useful when dealing with applications that use multiple threads. A **thread** is a code sequence that executes on its own. It will run at the same time as other threads within the same application. Concurrent threads do not pose a problem unless they are sharing data. When this happens it is possible for that data to become corrupted. The use of synchronized methods addresses this problem and prevents the data from being corrupted due to the interaction of the threads.

The use of synchronized methods includes some overhead. Thus, if the string is not being shared by multiple threads then the overhead introduced by the StringBuffer class is not needed. When synchronization is not needed, in most cases the StringBuilder class should be used.

> **Criteria for using string classes**
>
> If the string is not going to change, use the String class:
>
> - As it is immutable it is safe for sharing between multiple threads
> - The threads will only read them, which is normally a thread safe operation.
>
> If the string is going to change and it will be shared between threads, then use the StringBuffer class:
>
> - This class is designed for just this situation
> - Using this class in this situation will insure that the string is updated correctly
> - The chief drawback is that the methods may execute slower
>
> If the string is to change but will not be shared between the threads, use the StringBuilder class:
>
> - It allows modification of the strings but does not incur the overhead of synchronization
> - The methods of this class will execute as fast as, or faster, than those of the StringBuffer class

Unicode characters

Java uses the Unicode standard to define a character. However, this standard has evolved and changed, and Java has accommodated its changes. Originally the Unicode standard defined a character as a 2 byte 16-bit value which could be represented using printable characters or U+0000 through U+FFFF. Hexadecimal digits can be use to encode the Unicode characters whether they were printable or not.

However, the 2-byte encoding was not sufficient for all languages. So, version 4 of the Unicode standard introduced new characters above U+FFFF called **UTF-16 (16-bit Unicode Transformation Format)**. Java, in support of the new standard, uses the concept of **surrogate pairs**—pairs of 16-bit chars. These pairs are used to represent values from U+10000 through U+10FFFF. The leading or high value of the surrogate pair ranges from U+D800 through U+DBFF. The trailing or low value of the pair ranges from U+DC00 through U+DFFF. Characters in this latter range are called **supplementary characters**. These two special ranges are used to map any Unicode character into a surrogate pair. As of JDK 5.0, a character is represented using UTF-16.

The Character class

The Character class is a wrapper class for the char primitive data type. This data type supports the Unicode standard version 4.0. Characters are defined as fixed-width, 16-bit quantities.

The Character class – methods

The Character class possesses a number of methods for dealing with characters. Many of the Character methods are overloaded and can take either a char or a Unicode code point parameter. A code point is an abstraction used for a character and for our purposes is a Unicode character. The following table lists several Character methods that you are likely to encounter:

Methods	Description
isDigit	Returns true if the character is a digit
isLetter	Returns true if the character is a letter
isLetterOrDigit	Returns true if the character is a letter or a digit
isLowerCase	Returns true if the character is a lower case letter
isSpace	Returns true if the character is a space
isUpperCase	Returns true if the character is an upper case letter
toLowerCase	Returns the lower case equivalent of the character
toUpperCase	Returns the upper case equivalent of the character

The String class

The String class is a common class used to represent strings in Java. It is immutable which makes it thread safe. That is, multiple threads can access the same string and not worry about corrupting the string. Being immutable also means that it is of a fixed size.

One of the reasons the String class was made immutable was for security reasons. If a string is used to identify a resource that is protected, once permission has been granted for that resource, it may be possible to modify the string and then obtain access to another resource for which the user does not have permission. By making it immutable, this vulnerability is avoided.

While the String class is immutable, it may still appear to be mutable. Consider the following example:

```
String s = "Constant";
s = s + " and unchangeable";
System.out.println(s);
```

The output of this sequence is the string "Constant and unchangeable". As s is defined as a String type, the object referenced by the s identifier cannot change. When the second assignment statement is made, a new object is created that combines Constant and and unchangeable together to produce a new string Constant and unchangeable. Three String objects are created in the process:

- Constant
- and unchangeable
- Constant and unchangeable

The identifier, s, now references the new string Constant and unchangeable.

While we have access to these objects, we were not able to change them. We can access and read them but we cannot modify them.

We could have used the String class' concat method but this is not as straightforward:

```
s = "Constant";
s = s.concat(" and unchangeable");
System.out.println(s);
```

The following code illustrates several techniques for creating a String object. The first constructor will only produce an empty string. This is not of immediate value unless an empty immutable string located on the heap is needed in the application.

```
String firstString = new String();
String secondString = new String("The second string");
String thirdString = "The third string";
```

Java Data Types and Their Usage

In addition, there are two constructors that use the `StringBuffer` and `StringBuilder` classes. New `String` objects are created from these objects, as shown in the following code sequence:

```
StringBuffer stringBuffer =
   new StringBuffer("A StringBuffer string");
StringBuilder stringBuilder =
   new StringBuilder("A StringBuilder string");
String stringBufferBasedString = new String(stringBuffer);
String stringBuilderBasedString = new String(stringBuilder);
```

> Internally, a string of the `String` class is represented as an array of `char`.

String comparisons

String comparisons are not as straightforward as they might initially appear. If we wanted to compare two integers, we might use a statement such as the following:

```
if (count == max) {
  // Do something
}
```

However, for the comparisons of two strings, such as `s1` and `s2`, the following will often evaluate as `false`:

```
String s1 = "street";
String s2;

s2 = new String("street");

if (s1 == s2) {
  // False
}
```

The problem is that the variables `s1` and `s2` may be referencing different objects in memory. The if statement is comparing string reference variables and not the actual strings. As they are referencing different objects, the comparison returns `false`. It all depends on how the compiler and run-time system handles the strings internally.

When the `new` keyword is used, memory is allocated from the heap and assigned to the new object. In the case of a string literal however, this memory does not come from the heap but instead from a literal pool, or more specifically, the string intern pool. In Java, interned strings are placed into the permanent generation area of the JVM. This area also stores Java class declarations and class static variables, among other things.

String interning stores only one copy of each distinct string. This is to improve the execution of certain string methods and reduce the amount of space used to represent identical strings. The strings in this area are subject to garbage collection.

For example, if we create two string literals and a `String` object using the `new` keyword:

```java
String firstLiteral = "Albacore Tuna";
String secondLiteral = "Albacore Tuna";
String firstObject = new String("Albacore Tuna");

if(firstLiteral == secondLiteral) {
  System.out.println(
      "firstLiteral and secondLiteral are the same object");
} else {
  System.out.println(
      "firstLiteral and secondLiteral are not the same object");
}
if(firstLiteral == firstObject) {
  System.out.println(
      "firstLiteral and firstObject are the same object");
} else {
  System.out.println(
      "firstLiteral and firstObject are not the same object");
}
```

The output follows:

```
firstLiteral and secondLiteral are the same object
firstLiteral and firstObject are not the same object
```

The `String` class' `intern` method can be used to intern a string. Interning is performed automatically for all constant strings. When comparing interned strings, the equality operator can be used instead of having to use the `equals` method. This can save time for string intensive applications. It is easy to forget to intern a string, so be careful using the equality operator. In addition to this, the `intern` method can be an expensive method to use.

> Other objects besides the `String` type are interned by Java. These include wrapper objects and small integer values. Wrapper objects can result when a string concatenation operator is used with a primitive type. For more details visit http://docs.oracle.com/javase/specs/jls/se7/jls7.pdf and refer to sections 5.1.7 and 12.5.

To perform `String` comparisons, there is a number of `String` methods you can use including, but not necessarily limited to, the following:

Method	Purpose
equals	Compares two strings and returns `true` if they are equivalent
equalsIgnoreCase	Compares two strings while ignoring the case of the letters and returns `true` if they are equivalent
startsWith	Returns `true` if the string starts with the specified character sequence
endsWith	Returns `true` if the string ends with the specified character sequence
compareTo	Returns `-1` if the first string precedes the second, `0` if they are equal to each other, or `1` if the first string follows the second string

> Remember that strings start with an index `0`.

The following illustrates the use of various string comparisons:

```
String location = "Iceberg City";
if (location.equals("iceberg city"))
   System.out.println(location + " equals ' city'!");
else
   System.out.println(location +
     " does not equal 'iceberg city'");
```

```
    if (location.equals("Iceberg City"))
       System.out.println(location + " equals 'Iceberg City'!");
    else
       System.out.println(location +
          " does not equal 'Iceberg City'!");

    if (location.endsWith("City"))
       System.out.println(location + " ends with 'City'!");
    else
       System.out.println(location + " does not end with 'City'!");
```

The output is shown as follows:

```
Iceberg City does not equal 'iceberg city'
Iceberg City equals 'Iceberg City'!
Iceberg City ends with 'City'!
```

There are several things to consider when using this method. First, uppercase letters come before lowercase letters. This is the result of their ordering in Unicode. The same ordering applies to ASCII.

A string can have multiple internal representations. Many languages use the accent to differentiate or emphasize a character. For example, the French name, Irène, uses an accent and can be represented either as I r è n e or as the sequence I r e ` n e. The second sequence combines the e and ` to form the character è. If these two different internal representations were compared using the `equals` method, the method would return `false`. In this example, `\u0300` combines the grave accent with the letter e.

String firstIrene = "Irène";

```
    String secondIrene = "Ire\u0300ne";

    if (firstIrene.equals(secondIrene)) {
       System.out.println("The strings are equal.");
    } else {
       System.out.println("The strings are not equal.");
    }
```

The output of this code sequence is as follows:

```
The strings are not equal.
```

The `Collator` class can be used to manipulate strings in a locale-specific manner removing the problems of different internal string representations.

Basic string methods

There are several `String` methods you may encounter. These are illustrated in the following table:

Method	Purpose
length	Returns the length of the string.
charAt	Returns the position of a character given an index in the string.
substring	This method is overloaded and returns parts of the string.
indexOf	Returns the position of the first occurrence of a char or string.
lastIndexOf	Returns the position of the last occurrence of a char or string.

The following examples illustrate the use of these methods:

```
String sample = "catalog";
System.out.println(sample.length());
System.out.println(sample.charAt(0));
System.out.println(sample.charAt(sample.length()-1));
System.out.println(sample.substring(0,3));
System.out.println(sample.substring(4));
```

When this code is executed, we get the following output:

```
7
c
g
cat
log
```

Searching a string for a character or sequence of characters is a common requirement of many applications. The `indexOf` and `lastIndex` methods perform this type of operation:

```
String location = "Irene";
System.out.println(location.indexOf('I'));
System.out.println(location.lastIndexOf('e'));
System.out.println(location.indexOf('e'));
```

The results of these statements are as follows:

```
0
4
2
```

You can think of the position in a string as a position immediately before a character. These positions or indexes start at 0, as illustrated in the following diagram:

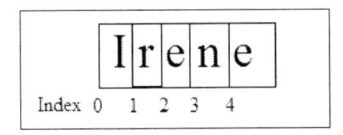

String length

The calculation of the length of a string can be a bit more complicated than suggested by the simple use of the `length` method. It depends upon what is being counted and how the string is represented internally.

Methods that can be used to determine the length of a string include:

- `length`: The standard method used
- `codePointCount`: This is used in conjunction with supplementary characters
- The `length` method of an array of bytes: This is used to determine the actual number of bytes used to hold the string

The actual length of the string in bytes can be important when storing a string. The amount of space allocated in a table of a database may need to be longer than simply the number of characters in the string.

Number/string conversions

The process of converting a number to a string is important. There are two approaches we can use. The first approach uses a static method, as shown in the following code sequence. The `valueOf` method will convert a number to a string:

```
String s1 = String.valueOf(304);
String s2 = String.valueOf(778.204);
```

The `intValue` and `doubleValue` methods take the object returned by the `valueOf` static method and return an integer or double number respectively:

```
int    num1 = Integer.valueOf("540").intValue();
double num2 = Double.valueOf("3.0654").doubleValue();
```

The second approach is to use the `parseInt` and `parseDouble` methods of their respective wrapper classes. Their use is illustrated as follows:

```
num1 = Integer.parseInt("540");
num2 = Double.parseDouble("3.0654");
```

Miscellaneous String methods

There are several miscellaneous methods that can be useful:

- `replace`: This replaces a part of a string with another string
- `toLowerCase`: This converts all characters to lower case in a string
- `toUpperCase`: This converts all characters to upper case in a string
- `trim`: This removes leading and trailing blanks

The use of these methods is illustrated below:

```
String oldString = " The gray fox ";
String newString;

newString = oldString.replace(' ','.');
System.out.println(newString);

newString = oldString.toLowerCase();
System.out.println(newString);

newString = oldString.toUpperCase();
System.out.println(newString);

newString = oldString.trim();
System.out.println("[" + newString +"]" );
```

The results are shown as follows:

```
.The.gray.fox.
 the gray fox
 THE GRAY FOX
[The gray fox]
```

The StringBuffer and StringBuilder classes

The `StringBuffer` and `StringBuilder` classes provide an alternative to the `String` class. Unlike the `String` class, they are mutable. This is sometimes helpful in making a program more efficiently. There are several commonly used methods that are available to manipulate a `StringBuffer` or a `StringBuilder` object. Several of these are illustrated in the example that follows. While the examples use the `StringBuffer` class, the `StringBuilder` method works the same way.

It is frequently necessary to append one string to another. This can be accomplished using the `append` method:

```
StringBuffer buffer = new StringBuffer();
buffer.append("World class");
buffer.append(" buffering mechanism!");
```

The following illustrates inserting a string into a buffer:

```
buffer.insert(6,"C");
```

A more detailed example:

```
StringBuffer buffer;
buffer = new StringBuffer();
buffer.append("World lass");
buffer.append(" buffering mechanism!");
buffer.insert(6,"C");
System.out.println(buffer.toString());
```

The result is as follows:

```
World Class buffering mechanism!
```

Summary

In this chapter we have examined how Java deals with the data. The use of the stack and heap are important programming concepts that go a long way in explaining concepts such as the scope and lifetime of a variable. The difference between objects and primitive data types was introduced along with the initialization of variables. The initialization process will be covered in more detail in *Chapter 6, Classes, Constructors, and Methods*. The operators available in Java were listed along with the rules of precedence and associativity. In addition, the manipulation of character and string data was presented.

In the next chapter we will explore the decision constructs available in Java and how they can be used effectively. This will build upon the data types presented here.

Certification objectives covered

In this chapter we covered the following:

- Understanding how Java handles data
- Investigating the relationships between identifiers, the Java class, and memory
- Defining the scope of the variables
- Initializing identifiers
- Building expressions using operators and operands
- Working with strings
- Understanding the difference between objects and primitive data types

Test your knowledge

1. What will happen when you compile and run the following code?

    ```
    public class ScopeClass{
        private int i = 35;
        public static void main(String argv[]){
            int i = 45;
            ScopeClass s = new ScopeClass ();
            s.someMethod();
        }
        public static void someMethod(){
            System.out.println(i);
        }
    }
    ```

    a. 35 will be printed out
    b. 45 will be printed out
    c. A compile time error will be generated
    d. An exception will be thrown

2. Which of the following lines will compile without warning or error?

    a. `char d="d";`
    b. `float f=3.1415;`
    c. `int i=34;`
    d. `byte b=257;`
    e. `boolean isPresent=true;`

3. Given the following declaration:
```
public class SomeClass{
    public int i;
    public static void main(String argv[]){
        SomeClass sc = new SomeClass();
        // Comment line
    }
}
```

Which of the following statements are correct if they replace the comment line?

    a. `System.out.println(i);`
    b. `System.out.println(sc.i);`
    c. `System.out.println(SomeClass.i);`
    d. `System.out.println((new SomeClass()).i);`

4. Given the following declaration:
```
StringBuilder sb = new StringBuilder;
```

Which of the following are valid uses of the sb variable?

    a. `sb.append(34.5);`
    b. `sb.deleteCharAt(34.5);`
    c. `sb.toInteger(3);`
    d. `sb.toString();`

5. Which of the following will return the position of the first letter a where the string s contains "banana"?

    a. `lastIndexOf(2,s);`
    b. `s.indexOf('a');`
    c. `s.charAt(2);`
    d. `indexOf(s,'v');`

6. Given the following code, which expression displays the word "Equal"?

```
String s1="Java";
String s2="java";
if(expression) {
   System.out.println("Equal");
} else {
   System.out.println("Not equal");
}
```

   a. s1==s2
   b. s1.matchCase(s2)
   c. s1.equalsIgnoreCase(s2)
   d. s1.equals(s2)

3
Decision Constructs

Every application makes some kind of decisions. In Java there are several programming constructs that we can use to make these decisions. These include logical expressions, the if statement, and the switch statement. The purpose of this chapter is to introduce these tools to you and illustrate how they can be used.

We will begin with a discussion of logical expressions as they are central to making decisions. Logical expressions are expressions that return a Boolean value.

Next, we will examine how the logical expressions can be used with the `if` statement and the conditional operator. There are numerous variations on how a `if` statement can be structured and we will look at their advantages and disadvantages.

This will be followed up with a discussion of the switch statement. Prior to Java 7, switch statements were based on integer or enumeration values. In Java 7, we can now use `String` values. The use of Strings and their potential pitfalls are examined.

The last section addresses general control structure issues along with the impact of floating point numbers when making decisions, comparing objects, and a discussion of useful ways of organizing code.

In this chapter we will:

- Examine the nature of decision constructs
- Examine the basics of logical expressions
- Learn how to use the `if` statement and look at its variations
- Understand the conditional operator and when it should be used
- Explore the switch statement and Java 7's use of strings with this statement
- Determine how floating point number comparisons impact control
- Examine the pitfalls associated with comparing objects

Flow of control

In any application, the flow of control within a program is determined by the order in which the statements are executed. It is convenient to consider groups of statements as blocks whose execution is controlled by decision statements. A block can be considered to be an individual statement or several statements contained within a block statement. A block statement in Java is a group of statements enclosed in open and close curly braces.

Control statement – an overview

Control structures are those parts of the language that determine the order in which individual statements are executed. Without control structures, statements are executed sequentially, as shown in the following code snippet:

```
hours ==35;
payRate = 8.55;
pay = hours * payRate;
System.out.println(pay);
```

To vary the order in which the statements are executed, control statements are used. In Java, these statements include:

- The `if` statement: This statement is frequently used to decide which of two branches to take

- Conditional operator: This statement is a simplified and limited form of the `if` statement

- The `switch` statement: This statement is used to decide which of several branches to take

The `switch` statement uses either an integer, enumeration, or a string value to make a decision. To understand the `if` statement requires an understanding of the logical expressions. This is covered in the next section.

Logical expressions

As with all expressions, a logical expression consists of operators and operands. In Java there are a limited number of logical operators as summarized in the following table. It is a subset of the operators listed in *Chapter 2, Java Data Types and Their Usage*:

Precedence	Operator	Associativity	Meaning
1	...		
	!	Right	Logical complement
	...		
...			
5	<, <=, >, and >=	Left	Logical
	`instanceof`	Left	Type comparison
6	== and !=	Left	Equality and inequality
...			
10	&&	Left	Logical AND
11	||	Left	Logical OR
12	?:	Right	Conditional
...			

The operands of a logical expression may be of any data type, but a logical expression will always evaluate to a `true` or a `false` value.

> Do not confuse the bitwise operators, &, ^, and | with the corresponding logical operators && and ||. The bitwise operators perform similar operations as the logical operators, but do it on a bit-by-bit basis.

Boolean variables

The words `true` and `false` are keywords in Java. Their names correspond to their values and they can be assigned to Boolean variables. A Boolean variable can be declared with the `boolean` keyword followed by the variable name and an optional initial value:

```
boolean isComplete;
boolean isReady = true;   // Initialized to true
boolean errorPresent;
```

When a logical expression is evaluated, it will return either a `true` or a `false` value. Examples of logical expressions include the following:

```
age > 45
age > 45 && departmentNumber == 200
((flowRate > minFlowRate) ||
    ((flowRate > maxFlowRate) && (valveA == off)))
```

It is a good practice to give a Boolean variable a name that reflects a state of `true` or `false`. The `isComplete` variable implies that an operation has completed. The variable, `isReady`, if set to true suggests that something is ready.

The equality operator

The equality operator consists of two equals signs and when evaluated will return either a `true` or a `false` value. The assignment operator uses a single equal sign and will modify its left operand. To illustrate these operators, consider the following example. If a value of a `rate` variable equals `100`, we could assume that an error is present. To reflect this error condition we could assign a `true` value to the `errorPresent` variable. This can be performed using both the assignment and the equality operators.

```
int rate;
rate = 100;
boolean errorPresent = rate==100;
System.out.println(errorPresent);
```

When the preceding code snippet is executed we get the following output:

true

The logical expression, `rate==100`, compares the value stored in `rate` to the integer literal `100`. If they are equal, which is the case here, the expression returns `true`. The `true` value is then assigned to `errorPresent`. If the value stored in `rate` had not been `100`, then the expression will return a value of `false`. We will examine the use of the equality operator in more depth in the *Comparing floating point numbers* and *Comparing objects* sections.

Relational operators

Relational operators are used to determine the relationship, or relative ordering, between two operands. These operators frequently use two symbols. For example, greater than or equal to is expressed using the `>=` operator. The ordering of the symbols is important. Using `=>` is not legal.

The relational operators are listed in the following table:

Operator	Meaning	Simple example
<	Less than	age<35
<=	Less than or equal	age<=35
>	Greater than	age>35
>=	Greater than or equal	age>=35
==	Equal	age==35

If we wish to determine whether an age is greater than 25 and less than 35, we will have to use the `age` variable twice and in combination with the `&&` operator, shown as follows:

```
age > 25 && age < 35
```

While the following expression may make sense to us, it is illegal in Java.

```
25 < age < 35
```

The reason that the variable `age` must be used twice, as in the preceding example, is because the relational operators are binary operators. That is, each binary operator acts on two operands. In the preceding expression, we compared 25 to see if it is less than `age`. The operation will return either a `true` or a `false` value. Next, the true or false result will be compared to 35 which does not make sense and is illegal.

These are the rules of the language. We can't break these rules, therefore it is important for us to understand the rules.

Logical operators

When we think about how we make decisions, we often use logical constructs such as AND and OR. We may make a decision if two conditions are both true, or we may decide to do something if either of the two conditions are true. The AND operator implies that both conditions must be true while OR implies that only one of the conditions need to be true.

These two operations are the basis for most logical expressions. We will often decide to do something if some condition is not true. We may decide not to walk the dog if it is raining. The NOT is also an operator used to make decisions. When used, it changes a true to a false and a false to a true.

Decision Constructs

There are three logical operators implementing these logical constructs in Java. They are summarized in the following table:

Operator	Meaning	Simple Example
&&	AND	age > 35 && height < 67
||	OR	age > 35 || height < 67
!	NOT	!(age > 35)

The AND, OR, and NOT operators are based on the following truth tables:

AND	True	False
True	True	False
False	False	False

OR	True	False
True	True	True
False	True	False

	Not
True	False
False	True

Some decisions can be more complicated and we express these decision evaluations using more complex combinations of the operators, &&, ||, or !. We may decide to go to the movie if it is not raining and if either we have enough money or a friend is going to pay our way.

If (it is not raining) AND

((we have enough money) OR (a friend will pay our way)) THEN

We will go to the movie

Parentheses can be used to control the order of evaluation of logical operators in the same way they control the order of evaluation of arithmetic operators. In the following code sequence, the existence of errors is determined by the values stored in the `rate` and `period` variables. These statements are equivalent, but differ in their use of parentheses. The use of the parentheses in the second statement is not strictly needed but it does make it clearer:

```
errorPresent = rate == 100 || period > 50;
errorPresent = (rate == 100) || (period > 50);
```

In the following statement, a set of parentheses is used to force the || operator to be executed before the && operator. As the && operator has higher precedence than the || operator, we need to use parentheses to change the order of evaluation:

```
errorPresent = ((period>50) || (rate==100)) && (yaw>56);
```

Parentheses always take precedence over other operators.

Short circuit evaluation

Short circuiting is the process of not completely evaluating a logical expression once the result becomes obvious. There are two operators in Java that short circuit—the logical && and || operators.

Using the && operator

Let's consider the logical && operator first. In the following example, we are trying to determine if sum is greater than 1200 and amount is less than 500. For the logical expression to return true, both conditions must be met:

```
if (sum > 1200 && amount <500)...
```

However, if the first condition is false then there is no reason to evaluate the rest of the expression. Regardless of the value of the second condition, the && operator will return false. With short circuiting, the second condition is not evaluated, saving some processing time especially if the operation is time consuming.

We can verify this behavior by using the following two functions. They both return false values and display messages as they execute:

```
private static boolean evaluateThis() {
    System.out.println("evaluateThis executed");
    return false;
}
private static boolean evaluateThat() {
    System.out.println("evaluateThat executed");
    return false;
}
```

Next, we use them in an if statement, shown as follows:

```
if(evaluateThis() && evaluateThat()) {
    System.out.println("The result is true");
} else {
    System.out.println("The result is false");
}
```

When we execute the preceding code sequence, we get the following output:

```
evaluateThis executed
The result is false
```

The evaluateThis method executed and returned false. As it returned false, the evaluateThat method was not executed.

Using the || operator

The logical || operator works in a similar fashion. If the first condition evaluates to `true`, there is no reason to evaluate the second condition. This is demonstrated in the following code sequence where the `evaluateThis` method has been modified to return `true`:

```
private static boolean evaluateThis() {
    System.out.println("evaluateThis executed");
    return true;
}

...

if(evaluateThis() || evaluateThat()) {
    System.out.println("The result is true");
} else {
    System.out.println("The result is false");
}
```

Executing this code sequence results in the following output:

```
evaluateThis executed
The result is true
```

Avoiding short circuit evaluation

Normally, short circuiting an expression is an efficient technique. However, if we invoked a method as we did in the last example and the program depended on the second method executing it can cause unanticipated problems. Suppose we had written the `evaluateThat` method as follows:

```
private static boolean evaluateThat() {
    System.out.println("evaluateThat executed");
    state = 10;
    return false;
}
```

When the logical expression is short circuited, the `state` variable will not be changed. If the programmer mistakenly assumed that the `evaluateThat` method would always be executed, then this could result in logic error when the value assigned to `state` is incorrect.

The `evaluateThat` method is said to have a side effect. One can argue whether or not it is a good practice to use methods that have side effects. Regardless, you may run across code that uses side effects and you need to understand its behavior.

One alternative to avoid the short circuit of logical expressions is to use the bitwise AND (`&`) and OR (`|`) operators. These bitwise operators perform the `&&` or `||` operations bit-by-bit for each bit of the operand. As the internal representation of the keywords, `true` and `false`, use a single bit, the result should be the same as returned by the corresponding logical operators. The difference is that the short circuit operation is not performed.

Using the previous example, if we use the `&` operator instead of the `&&` operator, as shown in the following code snippet:

```
if (evaluateThis() & evaluateThat()) {
    System.out.println("The result is true");
} else {
    System.out.println("The result is false");
}
```

We will get the following output, showing that both methods were executed when we execute the code:

```
evaluateThis executed
evaluateThat executed
The result is false
```

The if statement

The `if` statement is used to control the flow of execution based on a Boolean expression. There are two basic forms that can be used and there are several variations. The `if` statement consists of the `if` keyword, followed by a logical expression enclosed in parentheses and then by a statement. In the following diagram a graphical depiction of a simple `if` statement is presented:

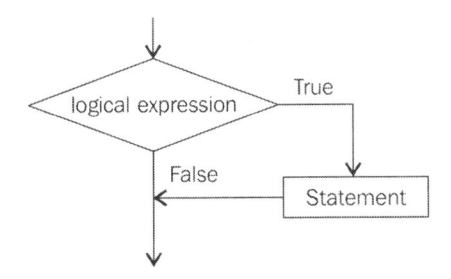

The following illustrates this form of the `if` statement where we compare `rate` to `100` and if it is equal to `100` we display a message to that effect:

```
if (rate==100) System.out.println("rate is equal to 100");
```

However, this is not as readable as the following equivalent example where we spilt the `if` statement across two lines:

```
if (rate == 100)
    System.out.println("rate is equal to 100");
```

As we will see later, it is better to always use a block statement with `if` statements. The following is logically equivalent to the previous `if` statement but is more readable and maintainable:

```
if (rate == 100) {
    System.out.println("rate is equal to 100");
}
```

The second form of the `if` statement uses the `else` keyword to designate the statement to execute, if the logical expression evaluates to `false`. The following diagram graphically illustrates this version of the `if` statement:

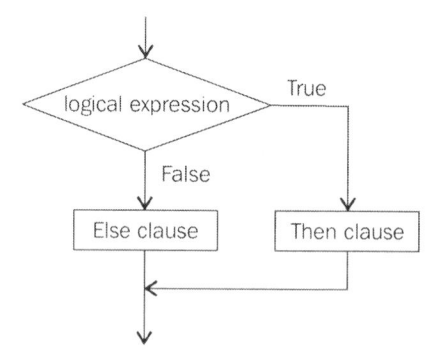

The `if` statement is illustrated as follows, using the preceding example:

```
if (rate == 100) {
   System.out.println("rate is equal to 100");
} else {
   System.out.println("rate is not equal to 100");
}
```

If the expression evaluates to `true`, the first block is executed and then control passes to the end of the `if` statement. If the expression evaluates to `false`, the second block is executed. In this example, each block consists of a single statement but it doesn't have to be this way. Multiple statements can be used within the blocks. The choice of the number of statements depends on what we are trying to do.

The simpler form of the `if` statement eliminates the `else` clause. Suppose we want to display an error message when a certain limit has been exceeded, otherwise do nothing. This can be accomplished by not using the `else` clause, as shown in the following code snippet:

```
if (amount > limit) {
   System.out.println("Your limit has been exceeded");
}
```

The only time that the message is displayed is when the limit has been exceeded. Note the use of the block statement. Even though it includes only a single statement it is still a good practice to use it. If we ever decide we need to do something more than simply displaying the error message, such as change the limit or reset the amount, then we will need a block statement. It is best to be prepared:

Some developers do not like this simpler form and will always use an else clause.

```
if (amount > limit) {
   System.out.println("Your limit has been exceeded");
} else {
   // Do nothing
}
```

Decision Constructs

The `Do nothing` comment serves to document the `else` clause. Should we decide to actually do something, such as place an order, then this is where we would add the code. With the use of the explicit `else` clause, we at least have to think about what could, or should, go there.

You may also encounter the **empty statement**. This statement consists of a single semicolon. When executed, it does nothing. It is commonly used as a place holder indicating that nothing is to be done. The previous `if` statement is modified in the following code snippet to use the empty statement:

```
if (amount > limit) {
   System.out.println("Your limit has been exceeded");
} else {
   ;   // Do nothing
}
```

This does not add anything to the `if` statement and it is not a problem to use it here. In *Chapter 5, Looping Constructs*, we will examine how the careless use of an empty statement can cause problems.

Nested if statements

Nesting `if` statements within each other provide another technique for making decisions. A `if` statement is nested if it is enclosed within the `then` or `else` clause of another `if` statement. In the following example, a `if` statement is found within the `then` clause of the first `if` statement:

```
if (limitIsNotExceeded) {
   System.out.println("Ready");
   if (variationIsAcceptable) {
      System.out.println(" to go!");
   } else {
      System.out.println(" - Not!");
   }
   // Additional processing
} else {
   System.out.println("Not Ok");
}
```

There is no limit on where the nested `if` can be used. It can be in either the `then` or `else` clauses. In addition, there is no limit as to how deeply they can be nested. We can place a `if` inside of a `if` inside of a `if` and so forth.

The else-if variation

In some programming languages there is a `elseif` keyword that provides a way of implementing a multiple-select `if` statement. Graphically, the logic of this statement is depicted in the following diagram:

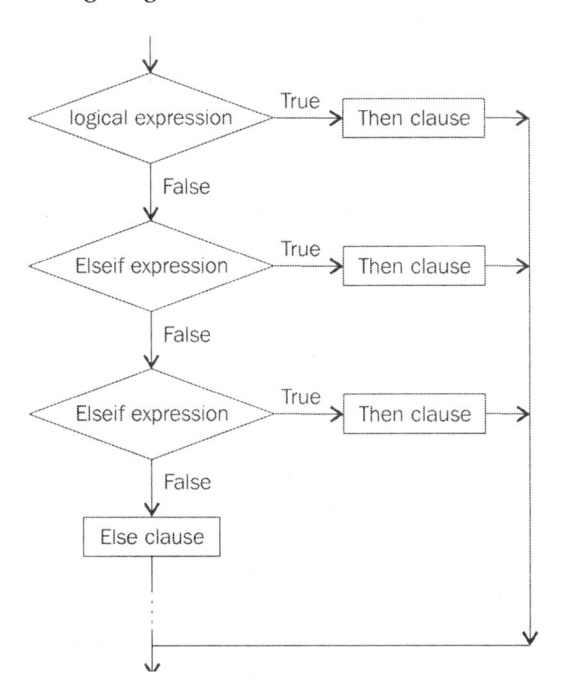

Java does not have the `elseif` keyword but the same effect can be achieved using nested if statements. Let's say we want to compute a shipping cost that is dependent on which of four regions of the country we are shipping to—East, North Central, South Central or West. We can do this using a series of `if` statements where each one is effectively nested inside of the `else` clause of the previous `if` statement. The first `if` statement that evaluates to true will have its body executed and the other `if` statements will be ignored:

```java
if (zone.equals("East")) {
    shippingCost = weight * 0.23f;
} else if (zone.equals("NorthCentral")) {
    shippingCost = weight * 0.35f;
} else if (zone.equals("SouthCentral")) {
    shippingCost = weight * 0.17f;
} else {
    shippingCost = weight * 0.25f;
}
```

Decision Constructs

This code sequence is equivalent to the following:

```
if (zone.equals("East")) {
   shippingCost = weight * 0.23f;
} else
   if (zone.equals("NorthCentral")) {
      shippingCost = weight * 0.35f;
   } else
      if (zone.equals("SouthCentral")) {
         shippingCost = weight * 0.17f;
      } else {
         shippingCost = weight * 0.25f;
      }
```

The second example achieves the same result as the first one but requires more indention. In the *The switch statement* section, we will demonstrate how to achieve the same result using a switch statement.

The if statement – usage issues

There are several issues that you should keep in mind when working with `if` statements. In this section we will examine the following issues:

- Misusing the equality operator
- Using Boolean variables instead of logical expressions
- Using true or false in a logical expression
- The perils of not using a block statement
- The dangling else problem

Misusing the equality operator

One nice feature of the Java language is the inability to write code that accidentally uses the assignment operator where the equality operator was meant to be. This happens frequently in the C programming language where the code compiles cleanly but results in logic errors or worse terminates abnormally at run-time.

For example, the following code snippet compares `rate` to see if it is equal to `100`:

```
if(rate == 100) {
   ...
}
```

However, if we had used the assignment operator instead, as shown in the following code snippet, we will generate a syntax error:

```
if(rate = 100) {
   ...
}
```

A syntax error similar to the following will be generated:

```
incompatible types
  required: boolean
  found:    int
```

This type of errors is eliminated in Java. The use of the equality operator with floating point numbers is covered in the *Comparing floating point numbers* section.

> Note that the error message says it found a `int` value. This is because the assignment operator returned a **residual value**. The assignment operator will modify the operand to its left and return the value that it assigned to that operand. This value is the residual value. It is left over from the operation.

Understanding the concept of residual value explains the error message. It also explains why the following expression works:

```
i = j = k = 10;
```

The effect of the expression is to assign `10` to each of the variables. The associativity for the assignment is right to left. That is, when there are multiple assignment operators in an expression, they are evaluated right to left. The value `10` is assigned to `k` and the assignment operator returned a residual value of `10`. The residual value is then assigned to `j` and so forth.

Using inverse operations

When using relational operators, there are frequently more ways than one to write the expression. For example, the following code sequence determines whether someone is of legal age or not:

```
final int LEGAL_AGE = 21;
int age = 12;

if(age >= LEGAL_AGE) {
   // Process
```

Decision Constructs

```
    } else {
        // Do not process
    }
```

However, this code sequence could have also been written as follows:

```
    if(age < LEGAL_AGE) {
        // Do not process
    } else {
        // Process
    }
```

Which approach is better? In this example, it could be argued that either approach will work. However, it is best to use the form which is most natural to the problem.

Note that the operations displayed in the following table are inverse operations:

Operation	Inverse Operation
<	>=
>	<=

Note the use of the constant, LEGAL_AGE. It is preferable to use an identifier when possible for values such as a legal age. If we did not and the value was used in multiple places, then changing the value only has to be done in one place. In addition, it avoids the mistake of accidentally using the wrong number in one of its occurrences. Also, making the number constant eliminates the possibility of accidentally modifying a value that should not be modified while the program is running.

Using Boolean variables instead of logical expressions

As we saw in the *Boolean variables* section, we can declare a Boolean variable and then use it as part of a logical expression. We can use a Boolean variable to hold the result of a logical expression, as shown in the following code snippet:

```
    boolean isLegalAge = age >= LEGAL_AGE;

    if (isLegalAge) {
        // Process
    } else {
        // Do not process
    }
```

This can be advantageous for two reasons:

- It allows us to re-use the result later, if we need to
- It makes the code more readable, if we use a meaningful Boolean variable name

We can also use the negation operator to change the order of the `then` and `else` clauses as follows:

```
if (!isLegalAge) {
    // Do not process
} else {
    // Process
}
```

This example will frequently be more confusing than the previous one. We could make it potentially even more confusing by using a poorly worded Boolean variable as follows:

```
if (!isNotLegalAge) {
    // Process
} else {
    // Do not process
}
```

While this is readable and valid, a general rule is to avoid double negatives just as we try to do in the English language.

Using true or false in a logical expression

The `true` and `false` keywords can be used in a logical expression. However, they are not necessary, are redundant, and clutter up the code with little value added. Note the use of the `true` keyword in the following logical `if` statement:

```
if (isLegalAge == true) {
    // Process
} else {
    // Do not process
}
```

The explicit use of the sub expression, `== true`, is not necessary. The same is true when using the `false` keyword. It is clearer and simpler to use the Boolean variable by itself as used in the earlier examples.

The perils of not using the block statement

As a block statement is considered to be a statement, this allows multiple statements to be included with either part of the `if` statement, as illustrated in the following code snippet:

```
if (isLegalAge) {
   System.out.println("Of legal age");
   System.out.println("Also of legal age");
} else {
   System.out.println("Not of legal age");
}
```

Block statements are not actually required when only one statement is needed for the `then` or `else` clause, but are encouraged. A similar, but invalid `if` statement, would look like this:

```
if (isLegalAge)
   System.out.println("Of legal age");
   System.out.println("Also of legal age");
else {
   System.out.println("Not of legal age");
}
```

The block statement was used to group the code together. The indention of the print statement does not group code. While it may imply that the first two `println` statements are part of the then portion of the `if` statement, the `if` statement will, in fact, result in a compile time error.

Here, the same code is presented but with different indention. The `if` statement has only a `if` clause with a single `println` statement. The second `println` statement follows and would be executed regardless of the value of the logical expression. This is followed by the else clause which is all by itself. The compiler treats this as a syntax error:

```
if (isLegalAge)
   System.out.println("Of legal age");
System.out.println("Also of legal age");
else {
   System.out.println("Not of legal age");
}
```

The generated syntax error will appear as follows:

```
'else' without 'if'
```

> A general rule of thumb is to always use block statements for the then and else parts of a if statement.

A more insidious problem can occur if an extra statement is in the else clause. Consider the following example:

```
if (isLegalAge)
    System.out.println("Of legal age");
else
    System.out.println("Not of legal age");
    System.out.println("Also not of legal age");
```

The third println statement is not a part of the else clause. Its indention is misleading. The equivalent code using proper indention is as follows:

```
if (isLegalAge)
    System.out.println("Of legal age");
else
    System.out.println("Not of legal age");
System.out.println("Also not of legal age");
```

It is clear that the third println statement will always be executed. The right way of writing this statement is as follows:

```
if (isLegalAge) {
    System.out.println("Of legal age");
} else {
    System.out.println("Not of legal age");
    System.out.println("Also not of legal age");
}
```

The dangling else problem

Another problem with not using a block statement is the dangling else problem. Consider the following series of tests where we need to make a few decisions:

- If limit is greater than 100 and the stateCode is equal to 45, we need to increase limit by 10
- If limit is not greater than 100, we need to decrease limit by 10

Decision Constructs

This logic is implemented below:

```
if (limit > 100)
    if (stateCode == 45)
        limit = limit+10;
else
    limit = limit-10;
```

However, this example does not properly implement the decision. There are at least two problems with this example. First, the indentation of the `else` keyword is irrelevant to the evaluation of the statements and is misleading. The `else` keyword is always paired with the closest `if` keyword which is, in this case, the second one. The compiler does not care how we indent our code. This means that the code is equivalent to the following:

```
if (limit > 100)
    if (stateCode == 45)
        limit = limit+10;
    else
        limit = limit-10;
```

Here, `stateCode` is only tested if the limit exceeds `100` and then `limit` is either increased or decreased by `10`.

Bear in mind that the compiler ignores whitespaces (blanks, tabs, new line, and so on) in any statement. The code sequence could be written without whitespace but this makes it harder to read:

```
if (limit > 100) if (stateCode == 45) limit = limit+10;
else limit = limit-10;
```

The second problem in this example is the failure to use block statements. Block statements not only provide a way of grouping statements but also provide a way of more clearly conveying the logic of an application. The problem can be resolved, as shown in the following code:

```
if (limit > 100) {
    if (stateCode == 45) {
        limit = limit+10;
    }
} else {
    limit = limit-10;
}
```

This is much clearer and accomplishes what was intended. It makes the debugging of the program easier and the code is more readable, which makes it more maintainable.

Conditional operator

The conditional operator is a condensed, limited form of a `if` statement. It is condensed in that the decision is limited to a single expression. It is limited because multiple statements cannot be included in the `then` or `else` clauses. It is sometimes called the **ternary operator** due to its three components.

The essential form of the operator is as follows:

LogicalExpression ? ThenExpression : ElseExpression

If the *LogicalExpression* evaluates to true, then the result of the *ThenExpression* is returned. Otherwise the result of the *ElseExpression* is returned.

The following simple example tests to see if a number is less than 10. If it is, 1 is returned, otherwise 2 is returned. The `then` and `else` expressions in the example are trivial integer literals.

```
result = (num < 10) ? 1 : 2;
```

This is equivalent to the following `if` statement:

```
if (num < 10) {
    result = 1;
} else {
    result = 2;
}
```

Consider the process for calculating overtime. If the employee works 40 hours or less, pay is computed as the number of hours worked times his pay rate. If more than 40 hours are worked, then the employee is paid time and a half for those hours over 40.

```
float hoursWorked;
float payRate;
float pay;

if (hoursWorked <= 40) {
    pay = hoursWorked * payRate;
} else {
    pay = 40 * payRate + (hoursWorked - 40) * payRate;
}
```

This operation can be perfomed using a conditional operator, shown as follows:

```
payRate = (hoursWorked <= 40) ?
    hoursWorked * payRate :
    40 * payRate + (hoursWorked - 40) * payRate;
```

Decision Constructs

While this solution is more compact, it is not as readable. In addition, the `then` and `else` clauses need to be an expression that returns some value. While the value does not have to be a number, it cannot be multiple statements unless a method is invoked containing those statements.

> The use of the conditional operator is discouraged, except in trivial cases, primarily due to its readability issues. It is usually more important to have readable, maintainable code than to save a few lines of code.

The switch statement

The purpose of a `switch` statement is to provide a convenient and simple method of making multi-branch selections based upon integer, enumeration, or `String` expression. The `switch` statement has the following basic form:

```
switch ( expression ) {
  //case clauses
}
```

There are usually multiple `case` clauses within the statement block. The basic form of the `case` clause uses the `case` keyword followed by a colon, zero or more statements, and usually a `break` statement. The `break` statement consists of a single keyword, `break`, shown as follows:

```
case <constant-expression>:
  //statements
break;
```

There is also an optional default clause that can be used. This will catch any values not caught by a `case` clause. This is demonstrated as follows:

```
default:
  //statements
break;   // Optional
```

The basic form of the `switch` statement is shown as follows:

```
switch (expression) {
  case value: statements
  case value: statements
  ...
  default: statements
}
```

Chapter 3

No two cases within a `switch` statement may have the same value. The `break` keyword is used to effectively end the code sequence and exit the `switch` statement.

When the expression is evaluated, control is passed to the case expression that matches the corresponding constant expression. If no case matches the value of the expression, control is passed to the `default` clause, if present. If the default prefix is not present, then none of the statements of `switch` will be executed.

We will illustrate the use of the `switch` statement for integer, enumeration, and `String` expressions. The use of strings in `switch` statements is new to Java 7.

Integer-based switch statements

The `if` statement can be used for choosing between multiple integer values. Consider the following example. A series of `if` statements could be used to calculate shipping cost based on an integer `zone` value, shown as follows:

```java
private static float computeShippingCost(
        int zone, float weight) {
    float shippingCost;

    if (zone == 5) {
        shippingCost = weight * 0.23f;
    } else if (zone == 6) {
        shippingCost = weight * 0.23f;
    } else if (zone == 15) {
        shippingCost = weight * 0.35f;
    } else if (zone == 18) {
        shippingCost = weight * 0.17f;
    } else {
        shippingCost = weight * 0.25f;
    }

    return shippingCost;
}
```

A `switch` statement could be used for the same purpose, shown as follows:

```java
switch (zone) {
    case 5:
        shippingCost = weight * 0.23f;
        break;
    case 6:
        shippingCost = weight * 0.23f;
        break;
```

Decision Constructs

```
   case 15:
      shippingCost = weight * 0.35f;
      break;
   case 18:
      shippingCost = weight * 0.17f;
      break;
   default:
      shippingCost = weight * 0.25f;
}
```

> Don't forget that integer data types include byte, char, short, and int. Any of these data types can be used with an integer switch statement. The data type long is not allowed.

The order of the case and default prefixes is unimportant. The only restriction is that the constant-expressions must all be unique. If the break statement is not the last case clause, then it may need a break statement, otherwise control will pass to the case clause that follows it:

```
switch (zone) {
   case 15:
      shippingCost = weight * 0.35f;
      break;
   default:
      shippingCost = weight * 0.25f;
      break; // Only needed if default is not
             // the last case clause
   case 5:
      shippingCost = weight * 0.23f;
      break;
   case 18:
      shippingCost = weight * 0.17f;
      break;
   case 6:
      shippingCost = weight * 0.23f;
      break;
}
```

> For readability purposes a natural order is usually maintained which is normally sequential. Using this order makes it easier to find a case clause and to make sure that cases are not left out accidentally.

[102]

The case and default prefixes do not alter the flow of control. Control will flow from one case to the next succeeding case unless the break statement is used. As zones 5 and 6 use the same formula to calculate the shipping cost, we could use back to back case statements without the use of the break statement:

```
switch (zone) {
   case 5:
   case 6:
      shippingCost = weight * 0.23f;
      break;
   case 15:
      shippingCost = weight * 0.35f;
      break;
   case 18:
      shippingCost = weight * 0.17f;
      break;
   default:
      shippingCost = weight * 0.25f;
}
```

Break statements are needed to insure that only those statements associated with a case are executed. Break is not necessarily needed at the end of the `default` clause as control will normally flow out of the `switch` statement. However, it is often included for purposes of completeness, and is necessary if the `default` clause is not the last case in the `switch` statement.

Enumeration-based switch statements

Enumerations can also be used with a `switch` statement. This can make it more readable and maintainable. The following is duplicated from *Chapter 2, Java Data Types and Their Usage*. The variable `direction` is used to control the behavior of the `switch` statement, shown as follows:

```
private static enum Directions {
    NORTH, SOUTH, EAST, WEST
};

Directions direction = Directions.NORTH;

switch (direction) {
   case NORTH:
      System.out.println("Going North");
      break;
   case SOUTH:
      System.out.println("Going South");
      break;
   case EAST:
```

```
            System.out.println("Going East");
            break;
        case WEST:
            System.out.println("Going West");
            break;
    }
```

When this is executed we get the following output:

```
Going North
```

String-based switch statements

To illustrate the use of a string with a `switch` statement, we will demonstrate the computation of the shipping cost based on a region as used in *The else-if variation* section. That implementation is shown as follows, for your convenience:

```
if (zone.equals("East")) {
   shippingCost = weight * 0.23f;
} else if (zone.equals("NorthCentral")) {
   shippingCost = weight * 0.35f;
} else if (zone.equals("SouthCentral")) {
   shippingCost = weight * 0.17f;
} else {
   shippingCost = weight * 0.25f;
}
```

Prior to Java 7, only integer variables could be used with a `switch` statement. By permitting the use of strings, programs can incorporate more readable code.

The following code snippet illustrates how to use a `String` variable with a `case` statement. The example provides an alternate implementation of the previous nested `if` statement:

```
switch (zone) {
   case "East":
      shippingCost = weight * 0.23f;
      break;
   case "NorthCentral":
      shippingCost = weight * 0.35f;
      break;
   case "SouthCentral":
      shippingCost = weight * 0.17f;
      break;
   default:
      shippingCost = weight * 0.25f;
}
```

String issues with the switch statement

There are two other issues that should be considered when using strings with a switch statement:

- When null values are encountered
- The case-sensitive nature of strings

When a null value has been assigned to a string variable used within a switch statement, a `java.lang.NullPointerException` exception will be thrown. Of course, this will happen whenever a method is executed against a reference variable that has been assigned a null value. In Java 7, there is additional support for handling null values found in the `java.util.Objects` class.

The second thing to remember about strings and the switch statement is that the comparison made within a `switch` statement is case-sensitive. In the previous example, if a string value of `east` had been used, the `East` case would not have been matched and the `default` case would have been executed.

Control structure issues

So far, we have identified several types of decision constructs that are available in Java. For example, simple decisions are easily handled using a `if` statement. Either-or type of decisions can be addressed using a `else if` clause or a `switch` statement.

The proper use of control structures is paramount in developing good code. However, there is more to making decisions than simply choosing between different control constructs. We also need to test our assumptions and handle unexpected situations.

In this section we will start by addressing a few general issues that you should keep in mind when using decision constructs. This will be followed by an examination of various floating point issues that can prove troublesome to those unfamiliar with floating point number limitations. Next, we will briefly introduce the topic of comparing objects and conclude with an overview of three basic coding activities that may prove helpful in understanding the nature of programming.

General decision constructs issues

There are several issues that are important in the use of decision constructs:

- The structure of the decision statements
- Testing your assumptions
- Planning for failure

Decision Constructs

The overall structure of a decision making process can be well structured or it can be an ad hoc sequence of statements that are hard to follow. A well organized approach to this structure can improve the readability and maintainability of the decision process.

A program may be well structured and yet may not work as intended. This is often due to invalid assumptions. For example, if the values for an age are assumed to be non-negative, then the code that is used may be well formed, and from a logical standpoint may be impeccable. However, if the assumption that good values for an age are used is wrong, then the results may not be as expected. For example, if the age of a person is entered as negative, then the logic may fail. It is important to always test your assumptions or at least make sure the underlying data has passed some sort of quality control check. Always expect the unexpected. Techniques to assist in this process include:

- Always keep a `else` clause
- Test your assumptions
- Throw exceptions (to be covered in *Chapter 8, Handling Exceptions in an Application*)
- Always use block statements

When all else fails, use debugging techniques.

Floating point number considerations

Floating point numbers are represented internally using the IEEE 754 Floating Point Arithmetic standard (`http://ieeexplore.ieee.org/xpl/mostRecentIssue.jsp?punumber=4610933`). These operations are normally performed in the software, because not all platforms provide hardware support for the standard. Performing these operations in the software will be slower than those executed directly in the hardware. The advantage of performing these operations in the software is that it supports the portability of applications.

Two floating point types are supported, `float` and `double`, with their precisions shown in the following table. In addition, the `Integer` and `Float` classes are wrapper classes for these two data types. Wrapper classes are used to encapsulate a value, such as an integer or floating point number:

Data type	Size (bytes)	Precision
float	4	23 binary digits
double	8	52 binary digits

Working with floating point numbers can be more complex than working with other data types. There are several aspects of floating point numbers that need to be considered. These include:

- Special floating point values
- Comparing floating point numbers
- Rounding errors

Special floating point values

There are several special floating point values as summarized in the following table. They exist so that when error conditions occur there will be a representation that can be used to identify the error.

These values exist so that error conditions such as arithmetic overflow, taking the square root of a negative number, and dividing by 0 can yield a result that can be represented within the floating point value without throwing an exception or otherwise terminating the application:

Value	Meaning	May be generated by
Not A Number	**NaN**: Represents the result of an operation that generated an undefined value	Division by zero
		Taking the square root of a negative number
Negative infinity	A very small value	A negative number divided by zero
Positive infinity	A very large value	A positive number divided by zero
Negative zero	Negative zero	A negative number is very close to zero but cannot be represented normally

NaN can be represented in code by `Float.NaN` and `Double.NaN`, if necessary. Performing an arithmetic operation with a NaN value will result in a NaN result. Casting a NaN to an integer will return 0 which could result in an application error. The use of NaN is illustrated in the following code sequence:

```
float num1 = 0.0f;

System.out.println(num1 / 0.0f);
System.out.println(Math.sqrt(-4));
System.out.println(Double.NaN + Double.NaN);
System.out.println(Float.NaN + 2);
System.out.println((int) Double.NaN);
```

Decision Constructs

When executed we get the following output:

```
NaN
NaN
NaN
NaN
0
```

Infinity is represented in Java using either of the following fields. As their names imply, we can represent either a negative or a positive infinity. Negative infinity implies a very small number and positive infinity represents a very large number:

- `Float.NEGATIVE_INFINITY`
- `Double.NEGATIVE_INFINITY`
- `Float.POSITIVE_INFINITY`
- `Double.POSITIVE_INFINITY`

In general, arithmetic operations involving infinite values will result in an infinite value. Those involving NaN will result in a NaN result. Division by zero will result in positive infinity. The following code snippet illustrates some of these operations:

```java
System.out.println(Float.NEGATIVE_INFINITY);
System.out.println(Double.NEGATIVE_INFINITY);
System.out.println(Float.POSITIVE_INFINITY);
System.out.println(Double.POSITIVE_INFINITY);
System.out.println(Float.POSITIVE_INFINITY+2);
System.out.println(1.0 / 0.0);
System.out.println((1.0 / 0.0) - (1.0 / 0.0));
System.out.println(23.0f / 0.0f);
System.out.println((int)(1.0 / 0.0));
System.out.println(
    Float.NEGATIVE_INFINITY == Double.NEGATIVE_INFINITY);
```

The output of this sequence is as follows:

```
-Infinity
-Infinity
Infinity
Infinity
Infinity
Infinity
NaN
Infinity
2147483647
True
```

A negative zero can be generated by dividing a negative number by positive infinity or a positive number divided by negative infinity, as illustrated in the following code snippet. The output of both statements will be a negative zero:

```
System.out.println(-1.0f / Float.POSITIVE_INFINITY);
System.out.println(1.0f / Float.NEGATIVE_INFINITY);
```

0 and -0 are distinct values. However, when compared to each other they will be determined to be equal to each other:

```
System.out.println(0 == -0);
```

This generates the following output:

True

Comparing floating point numbers

Floating point numbers, as represented in a computer, are not actually real numbers. That is, there is an infinite number of floating point numbers in the numbering system. However, either 32 or 64 bits are used to represent a floating point number. This means that only a finite number of floating point numbers can be represented exactly. For example, the fraction 1/3 cannot be represented exactly in base 10. If we try, we get something like 0.333333. Likewise, there are some floating point numbers that cannot be represented exactly in base 2 such as the fraction 1/10.

This implies that comparing floating point numbers can be difficult. Consider the following example where we divide two numbers and compare the result to the expected quotient of 6:

```
double num2 = 1.2f;
double num3 = 0.2f;
System.out.println((num2 / num3) == 6);
```

The result, when executed, gives us an unexpected value, as follows:

false

This is because these numbers are not represented exactly using the type `double`. To get around this problem, we can examine the result of the operation and see how much difference there is between what we expect and what we actually get. In the following sequence, a difference, `epsilon`, is defined as the maximum difference that is acceptable:

```
float epsilon = 0.000001f;
if (Math.abs((num2 / num3) - 6) < epsilon) {
    System.out.println("They are effectively equal");
```

```
        } else {
           System.out.println("They are not equal");
        }
```

When this is executed we get the following output:

`They are effectively equal`

Also, when comparing `Float` or `Double` objects using the `compareTo` method, remember that these objects are ordered as follows from low to high:

- Negative infinity
- Negative numbers
- -0.0
- 0.0
- Positive numbers
- Positive infinity
- NaN

For example, the following code will return `-1` indicating that a negative number is less than `-0.0`. The output will be `true`:

```
System.out.println((new Float(-2.0)).compareTo(-0.0f));
```

Rounding errors

It is important in some situations to watch out for rounding errors. Consider the following code sequence:

```
for(int i = 0; i < 10; i++) {
   sum += 0.1f;
}
System.out.println(sum);
```

When this code is executed, we get the following output:

`1.0000001`

This is the result of the rounding error whose origins derive from the inaccurate representation of the fraction 1/10.

> It is not a good idea to use floating point numbers for exact values. This is the case for dollars and cents. Instead, use `BigDecimal` as it provides better accuracy and is designed to support this type of operation.

The strictfp keyword

The `strictfp` keyword can be applied to a class, an interface, or a method. Prior to Java 2, all floating point calculations were performed in compliance with the IEEE 754 specifications. After Java 2, intermediate calculations were not restricted to the standard and allowed the use of extra bits available on some processors to improve precision. This can result in less portable applications due to differences in rounding. By using the `strictfp` keyword, all calculations will strictly abide by the IEEE standard.

Comparing objects

When comparing objects we need to consider:

- Comparing object references
- Comparing objects with the `equals` method

When comparing references, we determine whether two reference variables point to the same object or not. If we want to determine whether two reference variables that point to two different objects are the same, we use the `equals` method.

These two comparisons are illustrated in the following figure. The three references variables `r1`, `r2`, and `r3`, are used to reference two objects. The variables `r1` and `r2` reference Object 1 while `r3` references Object 2:

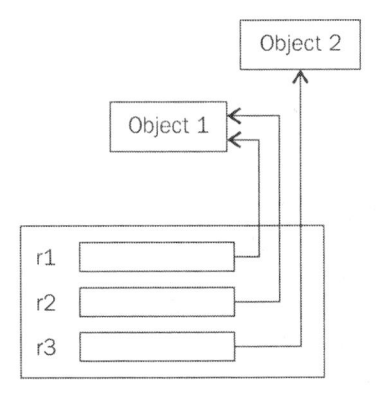

In this example, the following conditions are true:

- `r1 == r2`
- `r1 != r3`
- `r2 != r3`
- `r1.equals(r2)`

However, depending on the implementation of the `equals` method for the objects and the objects themselves, Object 1 may or may not be equivalent to Object 2. Comparisons of strings are covered in more detail in the *String comparisons* section in *Chapter 2, Java Data Types and Their Usage*. Overriding the `equals` method is discussed in *Chapter 6, Classes, Constructors, and Methods*.

Three basic coding activities

When writing code, it can be difficult to determine how to best organize your code. To help keep things in perspective, remember these three general coding activities:

- What you want to do
- How to do it
- When to do it

If a requirement of the application is to calculate the pay for an hourly employee, then:

- The "what" is to calculate pay
- The "how" determines how to write code to calculate the pay using the hours worked and the pay rate
- The "when" involves where to place the code, that is, after the hours worked and pay rate have been determined

While this may seem simple enough, many beginning programmers will have problems with the "when" of programming. This is especially true for event-driven programs typified by today's **Graphical User Interface (GUI)** based applications.

The goto statement

The `goto` statement is available in older programming languages and provides a powerful yet undisciplined way of transferring control within a program. Its use has often resulted in poorly organized programs and is discouraged. In Java, the use of the `goto` keyword is restricted. It simply cannot be used at all. It has been effectively banished from Java programming altogether.

However, statements with similar functionality to the goto statement still exist in many languages. For example, the break statement causes control to immediately be transferred to the end of the switch statement, and as we will see later, out of loops. Labels can also be used in conjunction with the break statement as we will see in the *Using labels* section in *Chapter 5, Looping Constructs*. This transfer is immediate and unconditional. It is effectively a goto statement. However, the break statement, and in similar fashion the return statement and exception handling, are considered to be more structured and safer. Control is not transferred to any arbitrary location within the program. It is only transferred to a specific location relative to statements at the end of the switch statement.

Summary

Decision making is an important aspect of programming. Most programs' utility is based on its ability to make certain decisions. The decision making process is based on the use of control constructs such as logical expressions, if statements, and switch statements.

There are different types of decisions to be made and are supported in Java with different control constructs. The primary ones discussed in this chapter included the if statement and the switch statement.

Care must be taken with the use of these statements to avoid the pitfalls possible with their use. These include misuse of the comparison operator, not using block statements as a matter of habit, and avoiding the dangling else problem. We also examined some of the issues that can occur when working with floating point numbers.

Decision making in Java can be simple or complex. Simple and complex either-or decisions are best handled using the if then else statement. For some of the simpler decisions, the simple if statement or conditional statement can be used.

Multiple choice decisions can be facilitated using either the if statement or the switch statement, depending on the nature of the decision. More complex decisions can be handled by the nesting of if statements and switch statements.

Now that we've learned about decision constructs, we are ready to examine how to use arrays and collections, which is the topic of the next chapter.

Certification objectives covered

With regards to certification objectives, we will examine:

- Using operators and decision constructs
- Using Java relational and logical operators
- Using parentheses to override operator precedence
- Creating if and if/else constructs
- Using the `switch` statement

Test your knowledge

1. What is the result of the following operation?

   `System.out.println(4 % 3);`

   a. 0
   b. 1
   c. 2
   d. 3

2. Which of the following expressions will evaluate to 7?

   a. `2 + 4 * 3- 7`
   b. `(2 + 4) * (3 - 7)`
   c. `2 + (4 * 3) - 7`
   d. `((2 + 4) * 3) - 7)`

3. What is the output of the following statement?

   `System.out.println( 16 >>> 3);`

   a. 1
   b. 2
   c. 4
   d. 8

4. Given the following declarations, which of the following if statements will compile without errors?

   ```
   int i = 3;
   int j = 3;
   int k = 3;
   ```

   a. if(i > j) {}
   b. if(i > j > k) {}
   c. if(i > j && i > k) {}
   d. if(i > j && > k) {}

5. What will be printed out when the following code is executed?

   ```
   switch (5) {
   case 0:
      System.out.println("zero");
      break;
   case 1:
      System.out.println("one");
   default:
      System.out.println("default");
   case 2:
      System.out.println("two");
   }
   ```

   a. one
   b. default and two
   c. one, two, and default
   d. Nothing, a compile-time error is generated

4
Using Arrays and Collections

This chapter, when boiled down to its essence, is about data structures. Specifically, it is about arrays—the `java.util.Arrays` and `java.util.ArrayList` classes. An array is a region of memory that can be addressed using a single variable name. It provides an efficient technique for accessing data in a sequential or random fashion. The `Arrays` class provides support for arrays while the `ArrayList` class provides array-like behavior but is not fixed in size.

We are concerned with how to create and use these data structures. A common operation is the traversal of an array or collection. We will see that Java supports several approaches permitting us to move through the elements of an array or an `ArrayList` object. Common to both arrays and collections is the ability to use the for-each statement. Iterators provide an alternate approach for accessing collections such as the `ArrayList`, and will also be discussed.

We will start by examining arrays in detail. This will include the creation and use of single and multidimensional arrays. Common array operations such as copying and sorting will be demonstrated.

As arrays are a simple data structure, most languages do not provide much support for operations on them. The `java.util.Arrays` class fills this void and supports important operations against arrays. These include copying, filling, and sorting the arrays.

The `java.util` package contains a number of interfaces and classes that can make working with collections of data easier. In this chapter we will examine the use of iterators and the `ArrayList` class which are part of this package. Iterators provide a technique for traversing over collections which can be very useful. The `ArrayList` class is frequently used instead of arrays when the size of the collection may change. It provides a number of valuable methods for the modification of collections.
We will also examine how to encapsulate a collection, such as the `ArrayList`, in another class.

Arrays

An array allows multiple values to be accessed using a single variable name. Each element of an array is of the same type. The element type can be a simple primitive data type or a reference to an object.

One-dimensional arrays are allocated to a contiguous region of memory. This means that the elements of the array can be accessed efficiently as they are adjacent to each other. Arrays use an integer index to access an element in the array. Indexes range from 0 to the length of the array minus one. We are able to access the elements of an array directly, in any order as needed by the application, without having to visit each element.

Though Java supports multidimensional arrays, one-dimensional arrays are most commonly used. Arrays can be used for a variety of purposes including:

- A collection of numbers representing ages
- A list of employee names
- A list of prices for an item in a store

The main disadvantage of arrays is that they have a fixed size. This makes it more difficult and less efficient to add, remove, or resize any list or collection of data represented by an array.

Our discussion will begin with the coverage of one-dimensional and multidimensional arrays. This is followed by discussions of common array techniques such as traversing an array and copying arrays. In the first two sections, we will use simple "for loops" to traverse array elements. Alternative methods are covered in the *Traversing arrays* section.

One-dimensional arrays

A one-dimensional array is intended to represent a simple linear list. The following code snippet illustrates the declaration and use of a one-dimensional array. The array, `ages`, is declared in the first statement as an array of the `int` type. In the second statement, memory is allocated for the array using the `new` operator. In this example, the array consists of 5 elements:

```
int[] ages;
ages = new int[5];
```

The `ages` array has 5 elements allocated to it. The first index of any array is 0. The largest index of an array is its length - 1. Thus, the last index of the array is 4. A runtime exception will be generated if an index used is outside the legal range of values for an array. The array could have been declared and created using a single statement, shown as follows:

```
int[] ages = new int[5];
```

As an array name is a reference to an array, it is possible to assign a different array to the variable later in the program. We will demonstrate this later in the discussion.

An array is an object that is allocated from an area of memory known as the **heap**. The heap and program stack were introduced in the *Stack and heap* section in *Chapter 2, Java Data Types and Their Usage*. In the following example, the first element of `ages` is assigned the value 35 and then displayed:

```
ages[0] = 35;
System.out.println(ages[0]);
```

Arrays possess the `length` property that returns the number of elements in the array. When the next code sequence is executed, it will return 5. Notice that `length` is not a method:

```
int length = ages.length;
System.out.println(length);
```

Arrays are represented in Java as objects. In the previous example, `ages` is an object reference variable which references the array that has been allocated to the heap. This is illustrated in the following diagram:

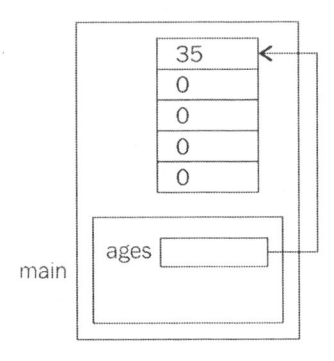

In this example, each element of the array was initialized to 0, by default, and then the first element was assigned a value of 35.

> Any attempt to use an index that is outside the bounds of the array will generate a `java.lang.ArrayIndexOutOfBoundsException` exception.

The placement of array brackets

There is a second option with regards to the placement of brackets when declaring an array. We can also place the brackets after the name of the array, shown as follows:

```
int ages[];
```

To the compiler, this is equivalent to the previous declaration. However, the placement of brackets with other uses of an array name is restricted. For example, we have to place the brackets after the array name when we are declaring or referencing elements of the array. If we attempt the following when declaring an array:

```
ages = new [5]int;
```

We will get a syntax error as follows:

`<identifier> expected`

`';' expected`

Likewise, if we try to use the brackets before the array name when referencing an element of the array, such as follows:

```
[0]ages = 0;
```

We get the following syntax error message:

`illegal start of expression`

`incompatible types`
`  required: int[]`
`  found:    int`

It is more common to see brackets used after the data type of the array. For example, most IDEs will at some point generate a `main` method. It frequently appears below with the brackets following the data type:

```
public static void main(String[] args) {
    ...
}
```

Also, consider the declaration:

```
int[] arr1, arr2;
```

Both `arr1` and `arr2` are declared as arrays. This is a simpler way of declaring more than one array on a single line. However, it can be argued that the following is a better format as it is more explicit:

```
int arr1[], arr2[];
```

It can be also argued that declaring more than one variable on a line is a bad form. The best way of declaring these two arrays is as follows:

```
int[] arr1;
int[] arr2;
```

Initializing arrays

The elements of an array are initialized to default values as shown in the following table. This table is duplicated from the *Initializing identifiers* section of *Chapter 2, Java Data Types and Their Usage*, for your convenience:

Data type	Default value (for fields)
boolean	false
byte	0
char	'\u0000'
short	0
int	0
long	0L
float	0.0f
double	0.0d
String (or any object)	null

In the previous example, we assigned a value of 35 to the first element of the array. This is a simple, yet tedious way of initializing the array to values other than the default value.

An alternate technique is to use a block statement to initialize the array. This is illustrated in the following example, where `ages` is initialized to five different values. It is not necessary to specify the array size when using the block statement to initialize an array:

```
int ages[] = {35, 10, 43, -5, 12};
```

Using Arrays and Collections

A syntax error will be generated if you try to specify the size of the array, shown as follows:

```
int ages[5] = {35, 10, 43, -5, 12};
```

The message will appear as follows:

`']' expected`

`';' expected`

If we want to display the content of an array, there are several methods available. Here, we will use simple indexes and the `length` property. In the *Traversing arrays* section we will demonstrate other techniques.

The following code sequence shows the difference between using the `toString` method and a for loop to display an array:

```
int ages[] = {35, 10, 43, -5, 12};
System.out.println(ages.toString());

for(int i = 0; i < ages.length; i++) {
    System.out.println(ages[i]);
}
```

When executed, we get the following output:

```
[I@27341e11
35
10
43
-5
12
```

Notice the use of the `toString` method does not return the contents of the array. Rather, it returns a strange representation of the array. We have no control over the string returned by the `toString` method. However, the for loop gives us what we expected.

> Remember that arrays in Java always begin with an index of 0.

Instead of hard coding the size of the array as 5, as we did in an earlier example, a better approach is to use a constant. For example, the entire sequence could be rewritten as follows:

```
static final int SIZE = 5;

int ages[] = new int[SIZE];
// initialize ages as needed

for(int i = 0; i < ages.length; i++) {
   System.out.println(ages[i]);
}
```

> Use named constants for array sizes. However, using the length attribute once the array is declared is preferred, as it is more maintainable should the array size change.

Arrays of objects

It is important to keep a clear distinction between an object reference variable and the object itself. An array of objects uses a reference variable, such as the names variable declared below, which is a single memory location that contains a reference to an array object. Each element of the array is another reference which may reference a string. Initially, they are assigned a null value:

```
public static void main(String args[]) {
   String names[] = new String[5];
   ...
}
```

The allocation of memory for this example is illustrated in the following diagram. However, we did not include the indexes for the array in the diagram. We can assume that the top element is at index 0 and the last element is at index 4:

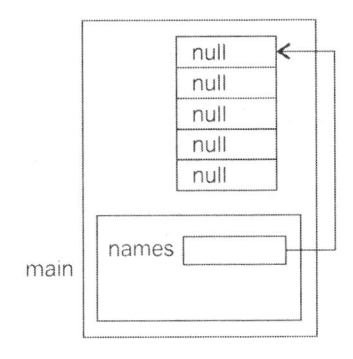

Using Arrays and Collections

When a string is assigned to an element of the array, the array element is modified to reference that string as illustrated as follows:

```
names[2] = "Steve";
```

The following diagram illustrates the modification of the reference at index 2 so that it references the string:

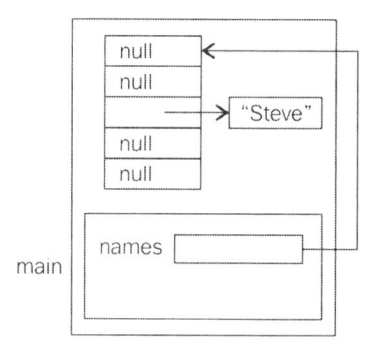

Be careful when using arrays that might contain null values. Consider the following code sequence where we display the contents of the `names` array:

```
for(int i = 0; i < names.length; i++) {
    System.out.println(names[i]);
}
```

When executed, we will get the following output:

null
null
Steve
null
null

Multidimensional arrays

Many applications require the use of arrays with two or more dimensions. Tabular data with rows and columns or one that uses the x/y coordinate system are good candidates for representation using a two-dimensional array. Three or more higher dimensions are not as common, but a coordinate system using an x, y, and z value would use three dimensions. In this section, we will demonstrate multidimensional arrays using integers. However, the techniques are equally applicable to arrays of objects.

An example of how a two-dimensional array is declared is shown as follows:

```
static final int ROWS = 2;
static final int COLS = 3;

int grades[][] = new int[ROWS][COLS];
```

This will create an array with 2 rows and 3 columns depicted logically, as shown in the following diagram:

Notice that the indexes start with a zero. We can initialize each element using a series of assignment statements as follows:

```
grades[0][0] = 0;
grades[0][1] = 1;
grades[0][2] = 2;
grades[1][0] = 3;
grades[1][1] = 4;
grades[1][2] = 5;
```

This is a bit tedious, but it illustrates the placement of numbers into the array, as shown in the following diagram:

Nested loops are useful for working with two-dimensional arrays. For example, to display the contents of such arrays, we will use a set of nested for loops as follows:

```
for (int rows = 0; rows < ROWS; rows++) {
   for (int cols = 0; cols < COLS; cols++) {
      System.out.printf("%d   ", grades[rows][cols]);
   }
   System.out.println();
}
```

Using Arrays and Collections

When executed, we get the output as follows:

```
0    1    2
3    4    5
```

Actually, Java does not support two-dimensional arrays in the strictest sense. In reality they are arrays of arrays. In languages such as C, two-dimensional arrays are stored in row-column order. This means that the two-dimensional array is mapped to a one-dimensional space where the first row is stored in the memory, followed by the second row and then the third and so forth. This is not applicable to Java.

Instead, what we actually have is a one-dimensional array of references to a series of other one-dimensional arrays. For example, we could have created the same `grades` array as:

```
grades = new int[ROWS][];
grades[0] = new int[COLS];
grades[1] = new int[COLS];
```

The array is allocated in the memory, as shown in the following diagram:

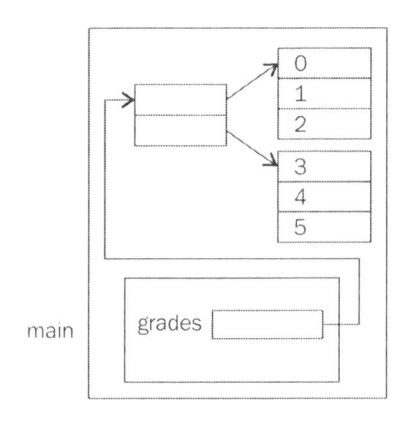

In a two-dimensional array, the rows do not necessarily have to be the same size. In the following code sequence, we create an array with different row lengths. Arrays of this type are called **ragged arrays**.

```
grades[0] = new int[4];
grades[1] = new int[2];
```

[126]

The memory allocation is similar to the previous example, except for the difference in array lengths, as shown in the following diagram:

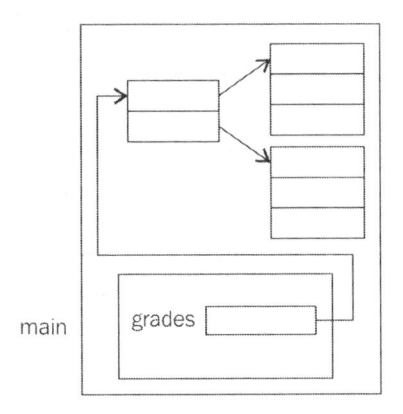

Array techniques

There are numerous techniques for working with arrays. In this section, we will examine many of these, including:

- Traversing arrays
- Comparing arrays
- Copying arrays
- Passing an array
- Using command-line arguments

We will demonstrate variations to each technique as appropriate. Passing a variable number of arguments to a method is covered in *Chapter 6, Classes, Constructors, and Methods*.

Traversing arrays

Traversing an array is the process of visiting each element of an array. This is frequently done starting with the first element and moving element by element until the end of the array is reached. However, it is also possible to move backwards or to skip elements. Here, we will focus on showing how we can traverse an array from beginning to end using two different techniques:

- Using simple for loops
- Using the for-each statement

Using Arrays and Collections

We will use the `ages` array, declared as follows, to illustrate how to traverse an array:

```
static final int SIZE = 5;
int[] ages = new int[SIZE];
```

In each example, we will use this code to initialize each element of the array to 5.

Using simple loops

Any simple loop can be used to traverse an array. Looping constructs are covered in more detail in *Chapter 5, Looping Constructs*. Here, we will use a for loop and a while loop. First, 'let's examine the for loop. In the following sequence, an integer variable starts at 0 and advances to the length of the array minus one:

```
for(int i = 0; i < ages.length; i++) {
    ages[i] = 5;
}
```

The equivalent while loop follows. Note the `i` variable is declared outside of the loop:

```
int i = 0;
while(i < ages.length) {
    ages[i++] = 5;
}
```

The for loop is generally preferable because we know the length of the array and it is simpler for these types of problems. For both examples, we used the `length` property of the array to control the loop. This is preferable to using a constant variable that may have been used to declare the array. Consider the following situation where we redefine the array:

```
int[] ages = new int[SIZE];
...
for(int i = 0; i < SIZE; i++) {
    ages[i] = 5;
}

// Array redefined
int[] ages = new int[DIFFERENT_SIZE];
...
for(int i = 0; i < SIZE; i++) {
    ages[i] = 5;
}
```

The second for loop will not execute properly because we forgot to change the `SIZE` constant and may even throw an exception if the array is smaller than `SIZE`. If we had used the `length` property instead, there would not have been a problem.

Notice, that the for loop, as written, declares the variable `i` within the for loop. This restricts access to the variable to only those statements within the for loop. In the while loop example, we declared `i` outside of the loop making it accessible inside, and outside, of the while loop. We could have rewritten the for loop to use an external `i` variable. However, it is considered to be better form to restrict access to a variable to only those statements that need access. Thus, if it is only needed inside of the loop then the for loop provides a better choice.

> Using simple for statements can result in off-by-one errors (starting at the wrong beginning or ending index). For example, if the value used as the last index is larger than the size of the array minus one, then a `ArrayIndexOutOfBoundsException` exception will be thrown.

Using the for-each statement

The for-each statement provides a more convenient method of traversing an array if we do not need explicit access to each element's index value. The for-each parentheses' body consists of a data type, a variable name, colon, and then an array (or collection). The statement will iterate through the array starting with the first element and ending with the last. During each iteration the variable references that array element. The following illustrates the use of this statement with the `ages` array. During the first iteration, `number` references `ages[0]`. During the second iteration, `number` references `ages[1]`. This continues for-each element of the array:

```
for(int number : ages) {
    number = 5;
}
```

The for-each statement makes it easy to traverse an array. However, if we need to use the index of an array element, the statement does not provide access to its value. The traditional for loop is needed to access the index.

The following table summarizes the differences between the use of the for loop and the for-each loop:

	for loop	for-each loop
Provides access to the array element	✓	✓
Provides access to the array index	✓	✗
Uses logical expression to control loop	✓	✗
Simplest	✗	✓

Comparing arrays

As an array variable is a reference variable, comparing array reference variables to determine equality will not always work. Here, we will examine several techniques for comparing arrays including:

- Element-by-element comparison
- Using the equality operator
- Using the `equals` method
- Using the `deepEquals` method

We will demonstrate these techniques by comparing two integer arrays. Consider the following example where two arrays, `arr1` and `arr2`, are equivalent after we initialize them to contain the same data:

```java
public static void main(String[] args) {
    int arr1[];
    int arr2[];
    arr1 = new int[5];
    arr2 = new int[5];

    for(int i = 0; i < 5; i++) {
       arr1[i] = 0;
       arr2[i] = 0;
    }
}
```

The following diagram shows how memory is allocated for both arrays:

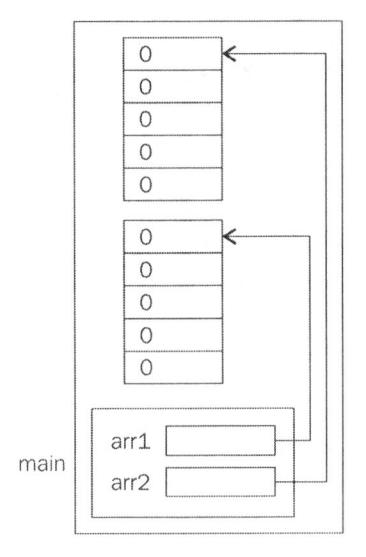

Element-by-element comparison

This simple approach will compare the corresponding elements of each array to determine if the arrays are equal. It starts by assuming they are equal and assigns a true value to the areEqual variable. If any comparison is false, then the variable is assigned the value of false:

```
boolean areEqual = true;
for (i = 0; i < 5; i++) {
   if(arr1[i] != arr2[i]) {
      areEqual = false;
   }
}
System.out.println(areEqual);
```

When this sequence is executed, it will display true. This is not the best approach. Using indexes is an error prone and tedious approach.

Using the equality operator

If we try to compare the two arrays using the equality operator, we find that the result of the comparison will be false:

```
System.out.println(arr1 == arr2);   //Displays false
```

Using Arrays and Collections

This is because we are comparing `arr1` and `arr2` which are array reference variables and not the arrays. The variables, `arr1` and `arr2`, reference different objects in memory. The contents of these two reference variables are different, therefore, when they are compared to each other they are not equal. They don't reference the same object.

Using the equals method

We can use the `equals` method with arrays as we can with other objects. In the following example, the output will be false even though they are equivalent. This is because the `equals` method, as applied to arrays, tests for object equivalency and not object value equivalency.

```
System.out.println(arr1.equals(arr2));   // Displays false
```

Object equivalency refers to the comparison of two object reference variables. If these variables reference the same object, they are considered to be equivalent. Object value equivalency refers to the condition where two distinct objects are considered to be equivalent because their internal values are the same.

Using the deepEquals method

To compare two arrays correctly we need to use the `Arrays` class' `equals` or `deepEquals` methods. The `equals` method performs a comparison using object identities. The `deepEquals` method performs a more in depth examination of the elements for value equivalency.

The following statement will display `true`:

```
System.out.println(Arrays.equals(arr1,arr2));
```

The `deepEquals` method requires an array of objects. The two-dimensional `grades` array, used in the *Multidimensional arrays* section, satisfies the requirement as it is an array of arrays, that is, an array that references other arrays (which are objects).

If we create a second grade array, `grades2`, and populate it with the same values as `grades`, we can use these methods to test for equality. The creation and initialization of the `grades2` array follows:

```
int grades2[][];
grades2 = new int[ROWS][];
grades2[0] = new int[COLS];
grades2[1] = new int[COLS];

grades2[0][0] = 0;
grades2[0][1] = 1;
grades2[0][2] = 2;
```

```
grades2[1][0] = 3;
grades2[1][1] = 4;
grades2[1][2] = 5;
```

If we execute the following sequence:

```
System.out.println(grades == grades2);
System.out.println(grades.equals(grades2));
System.out.println(Arrays.equals(grades, grades2));
System.out.println(Arrays.deepEquals(grades, grades2));
```

We will get the following output:

`false`

`false`

`false`

`true`

The first three comparisons returned `false` because they did not adequately compare the two arrays. The fourth technique compared the arrays in depth and accurately determined their equivalency.

The following table summarizes these techniques:

Technique	Comment
Element-by-element comparison	This will compare arrays properly, if implemented correctly.
Using the equality operator	This only works properly if the two reference variables reference the same object.
Using the array's `equals` method	This only works properly if the two reference variables reference the same object.
Using the `Array`'s class `equals` method	This will work for one-dimensional arrays.
Using the `Array`'s class `deepEquals` method	This performs a deeper comparison using the object's `equals` method.

Copying arrays

There are times when we need to copy one array to another. In this section, we will examine various techniques to achieve this goal. These include:

- Simple element-by-element copy
- Using the `System.arraycopy` method
- Using the `Arrays.copyOf` method

- Using the `Arrays.copyOfRange` method
- Using the `clone` method

We will demonstrate the techniques using two one-dimensional arrays as declared below:

```
int arr1[] = new int[5];
int arr2[] = new int[5];
```

We will initialize each element of `arr1` to its index with the following code:

```
for(int i = 0; i < arr1.length; i++) {
    arr1[i] = i;
}
```

In this section's examples, the content of the destination array follows as a comment.

We will also use the terms, **shallow copy** and **deep copy**. Shallow copy refers to when only the reference values are copied. After the copy operation, the original object has not been duplicated. In a deep copy, the reference to the object is not copied. Instead, a new copy of the object is created. We will see how some of the techniques illustrated here only perform a shallow copy which may not always be desirable.

Simple element-by-element copy

A simple technique is to use a for loop as illustrated below:

```
for(int i = 0; i < arr1.length; i++) {
    arr2[i] = arr1[i];
}
// 0, 1, 2, 3, 4
```

This is a simple approach but you need to be careful to use the correct array indexes. This technique becomes more complicated with multidimensional arrays.

Using the System.arraycopy method

The `System` class' `arraycopy` method will attempt to copy all, or part, of one array to another. The beginning position in each array is specified, along with the number of elements to copy.

To copy all of the elements of `arr1` to `arr2` we can use the following code:

```
System.arraycopy(arr1, 0, arr2, 0, 5);
// 0, 1, 2, 3, 4
```

The parameters of this method are detailed in the following table:

Parameter	Description
1	The source array
2	The starting index in the source array
3	The destination array
4	The starting index in the destination array
5	The number of elements to copy

The next sequence copies the first three elements of arr1 to the last three elements of arr2:

```
System.arraycopy(arr1, 0, arr2, 2, 3);
// 0 0 0 1 2
```

We can also copy part of one array to other positions within the same array. Here we copy the first two elements to the last two elements of the arr1 array:

```
System.arraycopy(arr1, 0, arr1, 3, 2);
// 0 1 2 0 1
```

There are numerous opportunities for exceptions to occur when using this technique. If either array reference is null, a NullPointerException exception is thrown. If the array indexes are invalid, then we will get a IndexOutOfBoundsException exception.

The arraycopy method will copy the specified elements of the source array to the corresponding element of the destination array. There are two possible results depending on the data type of the array. They are as follows:

- If the array element type is a primitive data type, then the two arrays are effectively identical.
- If the array element type is a reference, then both arrays will be identical but they will both reference the same objects. This is usually not the effect anticipated or desired.

In the following code sequence, an attempt is made to create an identical copy of the StringBuilder array, arr3:

```
StringBuilder arr3[] = new StringBuilder[4];
arr3[0] = new StringBuilder("Pine");
arr3[1] = new StringBuilder("Oak");
arr3[2] = new StringBuilder("Maple");
arr3[3] = new StringBuilder("Walnut");
```

```
StringBuilder arr4[] = new StringBuilder[4];
System.arraycopy(arr3, 0, arr4, 0, 4);
```

However, `arr4` contains the same object reference variables used by `arr3`. The corresponding element of both arrays reference the same object. The creation of an identical array with references to distinct strings is achieved with the following code:

```
for (int i = 0; i < arr3.length; i++) {
   arr4[i] = new StringBuilder(arr3[i]);
}
```

We created a new `StringBuilder` object for-each element of the destination array. This approach is necessary if a deep copy is needed.

Using the Arrays.copyOf method

The `Arrays` class' `copyOf` method will create a new array based on an existing array. The first argument of the method specifies the original array. Its second argument specifies how many elements to copy. In the following example, we create a new array based on the first three elements of `arr1`:

```
arr2 = Arrays.copyOf(arr1, 3);
// 0  1  2
```

The new array can be larger than the original array as illustrated with the following code:

```
arr2 = Arrays.copyOf(arr1, 10);
// 0  1  2  3  4  0  0  0  0  0
```

The last five elements of `arr2` will be padded with zeros.

If the array is an array of objects, a copy of the original object is assigned to the new array.

Using the Arrays.copyOfRange method

The `Arrays` class' `copyOfRange` method will create a new array based on a sub-range of elements in an existing array. The first argument of the method specifies the original array. Its second argument specifies the beginning index and the last argument specifies the ending index exclusive. In the following example, we create a new array based on the last two elements of `arr1`:

```
arr2 = Arrays.copyOfRange(arr1, 3, 5);
//  3  4
```

Notice that the last argument is not a valid index for the `arr1` array. This works here because the last argument is exclusive. It does not include that element.

In fact, if we specify a value such as 8 in the next example, the new array is padded with zeros:

```
arr2 = Arrays.copyOfRange(arr1, 3, 8);
//            3  4  0  0  0
```

Using the clone method

You can also use the `Object` class' `clone` method to create a copy of an array:

```
arr2 = arr1.clone();
```

However, this only makes a shallow copy of the original object. With an array of primitives such as the above integer array, this is not a problem. With an array of references to objects, both arrays will reference the same objects.

The following table summarizes the copying techniques introduced in this section:

Technique	Comment
Simple element-by-element copy	Tedious but can implement either a shallow or deep copy
Using the `System.arraycopy` method	Performs a shallow copy
Using the `Arrays.copyOf` method	Performs a deep copy of the entire array
Using the `Arrays.copyOfRange` method	Performs a deep copy of part of an array
Using the `clone` method	Performs a shallow copy

Passing arrays

The advantage of passing an array to a method is that it allows us to perform the same operation against more than one array. To pass an array to a method, we use the array name in the method call and declare a reference to the passed array in the method. This is illustrated below with a call to the `displayArray` method. This method simply displays the array.

```
displayArray(arr2);
    ...
private static void displayArray(int arr[]) {
    for(int number : arr) {
        System.out.print(number + " ");
    }
    System.out.println();
}
```

Using Arrays and Collections

Notice that we are "passing a reference" to the `arr2` array "by value". That is, if we want, we can read and write the elements of the `arr2` array in the method. However, if we modify the `arr` parameter, the original `arr2` variable is not modified.

Consider the method in the following code that attempts to change what the `arr2` reference variable points to:

```
System.out.println("Length of arr2: " + arr2.length);
changeArray(arr2);
System.out.println("Length of arr2: " + arr2.length);
...
private static void changeArray(int arr[]) {
   arr = new int[100];
   System.out.println("Length of arr: " + arr.length);
}
```

When we execute this code, we get the following output:

```
Length of arr2: 5
Length of arr: 100
Length of arr2: 5
```

The value of `arr` was changed but the value of `arr2` was not changed. The following diagram should help clarify this behavior:

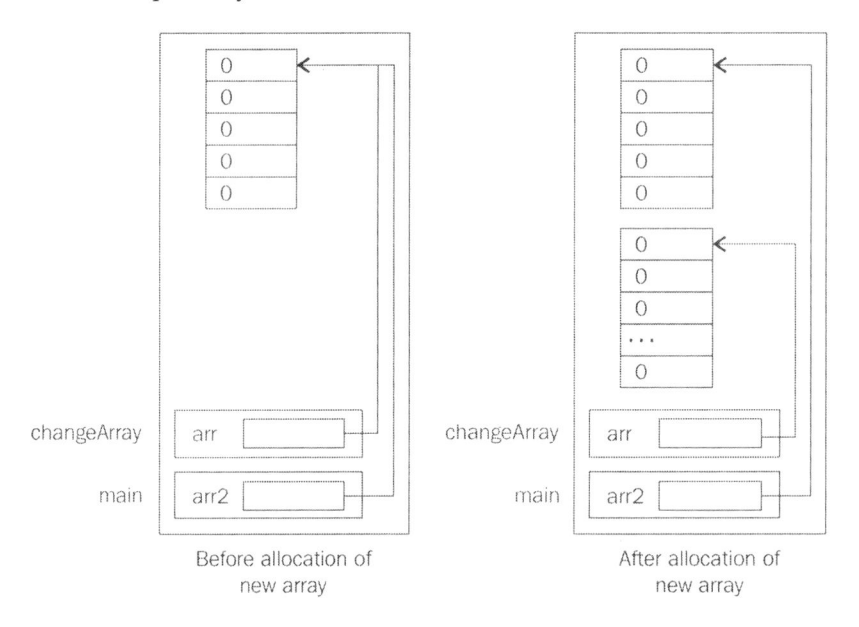

Before allocation of new array

After allocation of new array

Using command-line arguments

When a Java application executes, the first method that is executed is the `main` method. This method passes an argument, an array of `String` objects called `args`. These strings correspond to those provided on the command line.

The `length` property of a Java array will tell us how many command-line arguments were used. The first argument of the array will contain the first command-line parameter. The second will contain the second command-line parameter, and so forth.

The following `CommandLineDemo` application illustrates the use of the `args` array:

```java
public class CommandLineDemo {

  public static void main(String args[]) {
     System.out.println("The command line has " +
        args.length + " arguments");
     for (int i = 0; i < args.length; i++) {
        System.out.println("\tArgument Number " + i +
             ": " + args[i]);
     }
  }
}
```

Consider that the application is invoked with the following command-line arguments:

`java CommandLineDemo /D 1024 /f test.dat`

The output of the program would appear as follows:

```
The command line has 4 arguments
        Argument Number 0: /D
        Argument Number 1: 1024
        Argument Number 2: /f
        Argument Number 3: test.dat
```

The Arrays class

The `java.util.Arrays` class possesses several methods useful for working with arrays. Every method of the class is a static method which means that we do not have to create an instance of the `Arrays` class before we use its methods. The class is designed to work with arrays and perform common operations on arrays. The types of operations available include:

- Returning a `List` based on an array
- Performing a binary search
- Making copies of an array
- Determining the equality of two arrays
- Filling arrays
- Sorting arrays

We have seen the use of several of these techniques in earlier sections. Here we will demonstrate the use of the `asList`, `fill`, `toString`, and `deepToString` methods.

Consider the following declarations. We will declare an integer array and then an array list. Two strings will be added to the `ArrayList` object. We will also create an array of mixed objects and an array of strings. The `ArrayList` class is discussed in more detail in the *ArrayList* section:

```
int arr1[] = new int[5];
ArrayList list = new ArrayList();
list.add("item 1");
list.add("item 2");

Object arr2[] = {"item 3", new Integer(5), list};
String arr3[] = {"Pine", "Oak", "Maple", "Walnut"};
```

Next, we will fill the integer array with the number 5 using the `fill` method:

```
Arrays.fill(arr1,5);
```

The `asList`, `toString`, and `deepToString` methods are then used against these arrays, shown as follows:

```
System.out.println(Arrays.asList(arr3));
System.out.println(Arrays.toString(arr1));
System.out.println(Arrays.deepToString(arr2));
```

When executed we get the following output:

```
[Pine, Oak, Maple, Walnut]
 [5, 5, 5, 5, 5]
[item 3, 5, [item 1, item 2]]
```

The `asList` method takes its array argument and returns a `java.util.List` object representing the array. If either the array or the list is modified, their corresponding elements are modified. This is demonstrated in the following example:

```
List list2 = Arrays.asList(arr3);
list2.set(0, "Birch");
System.out.println(Arrays.toString(arr3));
```

The output of this sequence follows:

`[Birch, Oak, Maple, Walnut]`

The `toString` method returns a string representation of the array. The `deepToString` method is intended to return a string representation of its array argument where the array is more complex. This was reflected in `arr2` which contains different objects including a list.

Key points to remember when using arrays

When working with arrays remember:

- Array indexes start at 0
- Indexes have to be integers
- An array can hold primitive data types or objects
- Arrays provide constant time random access which is an efficient way of accessing data
- Arrays provide good locality of reference
- Arrays are more difficult to insert or remove elements than other data structures
- An index to an invalid element is possible

Locality of reference refers to the idea that if one data item is accessed, it is likely that another nearby data item will also be accessed. This results in faster read and write operations and is an important concept in virtual operating systems. Accessing elements of an array can be faster than accessing elements of a linked list when the linked list is spread across the memory.

Be careful when accessing elements of an array. If the array is not properly initialized, then the element being indexed may be invalid resulting in a run-time or logic error.

Collections

The Collections Framework was introduced in Java 2 as a set of interfaces and classes that are superior to many of the interfaces and classes found in the earlier `java.util` package such as `Vector`, `Stack`, and `HashTable`. These interfaces and classes should always be used instead of the older ones whenever possible. Many of the Collection Framework interfaces and classes are summarized in the following table:

Interface	Class
Set	HashSet
	TreeSet
List	ArrayList
	LinkedList
Map	HashMap
	TreeMap

The Collection Framework is covered in more detail at http://java.sun.com/developer/onlineTraining/collections/Collection.html. Here, we will address the `ArrayList` class as it is a certification topic. It is recommended that the `ArrayList` class be used when a `List` is needed. As we will see, iterators are used with the `ArrayList` to support traversal of the list. We will start our discussion with coverage of this topic.

Iterators

Iterators provide a means of traversing a set of data. It can be used with arrays and various classes in the Collection Framework. The `Iterator` interface supports the following methods:

- `next`: This method returns the next element
- `hasNext`: This method returns `true` if there are additional elements
- `remove`: This method removes the element from the list

The `remove` method is an optional `Iterator` method. If an attempt is made to use this method and the implementation of the interface does not support this method, then an `UnsupportedOperationException` exception is thrown.

The `ListIterator` interface, when available, is an alternative to the `Iterator` interface. It uses the same methods and provides additional capabilities including:

- Traversal of the list in either direction
- Modification of its elements
- Access to the element's position

The methods of the `ListIterator` interface include the following:

- `next`: This method returns the next element
- `previous`: This method returns the previous element
- `hasNext`: This method returns `true` if there are additional elements that follow the current one
- `hasPrevious`: This method returns `true` if there are additional elements that precede the current one
- `nextIndex`: This method returns the index of the next element to be returned by the `next` method
- `previousIndex`: This method returns the index of the previous element to be returned by the `previous` method
- `add`: This method inserts an element into the list (optional)
- `remove`: This method removes the element from the list (optional)
- `set`: This method replaces an element in the list (optional)

ArrayList

The `ArrayList` class has several useful characteristics:

- It is flexible
- Grows as needed
- Possesses many useful methods
- Access is performed in constant time
- Insertion/deletion is performed in linear time
- Can be traversed with indexes, for-each loops, or iterators

`ArrayList` uses an array internally. When it needs to grow, elements are copied from the old array to the new array.

The `ArrayList` class is not synchronized. When an iterator is obtained for a `ArrayList` object, it is susceptible to possible simultaneous overwrites with loss of data if modified in a concurrent fashion. When multiple threads access the same object, it is possible that they may all write to the object at the same time, that is, concurrently. When this simultaneous overwrite occurs, a `ConcurrentModificationException` exception is thrown.

Creating ArrayList

The `ArrayList` class possesses the following three constructors:

- A default constructor
- One that accepts a `Collection` object
- One that accepts an initial capacity

The capacity of a `ArrayList` object refers to how many elements the list can hold. When more elements need to be added and the list is full, the size of the list will be automatically increased. The initial capacity of a `ArrayList` created with its default constructor is 10. The following example creates two lists, one with a capacity of 10 and the second with a capacity of 20:

```
ArrayList list1 = new ArrayList();
ArrayList list2 = new ArrayList(20);
```

The `ArrayList` class supports generics. Here, a list of strings is created:

```
ArrayList<String> list3 = new ArrayList<String>();
```

We will use `list3` in the examples that follow.

Adding elements

There are several methods available for adding elements to an `ArrayList`. They can be placed into one of the following two categories:

- Appends one or more elements to the end of the list
- Inserts one or more elements at a position within the list

The simplest case is illustrated here where a string is added to the end of `creatures`:

```
ArrayList<String> creatures = new ArrayList<String>();
creatures.add("Mutant");
creatures.add("Alien");
creatures.add("Zombie");
System.out.println(creatures);
```

The output of the print statement follows:

`[Mutant, Alien, Zombie]`

To insert an element at the index after the first element we use an index of 1:

```
creatures.add(1,"Godzilla");
System.out.println(creatures);
```

Executing the code will verify the actions, as shown below:

`[Mutant, Godzilla, Alien, Zombie]`

The `addAll` method can also be used with `Collections`, as illustrated below:

```
ArrayList<String> cuddles = new ArrayList<String>();
cuddles.add("Tribbles");
cuddles.add("Ewoks");

creatures.addAll(2, cuddles);
System.out.println(creatures);
```

This will result in the `cuddles` being placed after the second element in the list, as shown below:

`[Mutant, Godzilla, Tribbles, Ewoks, Alien, Zombie]`

The `addAll` method can also be used without an index argument. In this case, the new elements are added to the end of the list.

Retrieving elements

To retrieve an element at a given position, use the `get` method. This method takes a single integer index value. In the following example, we retrieve the third element of the list. Assuming that the creatures list contains [Mutant, Godzilla, Tribbles, Ewoks, Alien, Zombie], the following statement will retrieve `Tribbles`:

```
String element = creatures.get(2);
```

Using Arrays and Collections

The index of an element can be obtained using the `indexOf` method as illustrated in the next code sequence. If the element does not exist, the method will return a -1.

```
System.out.println(creatures.indexOf("Tribbles"));
System.out.println(creatures.indexOf("King Kong"));
```

Executing this code will generate the following output:

```
2
-1
```

The `indexOf` method will return the index of the first element found. The `lastIndexOf` method will return the index of the last element found in the list.

The `toArray` method will return an array of the objects in the list. In this example, the `creatures` list is returned and assigned to the `complete` array. If the array is not large enough, as is the case here, a new array is created and returned.

```
String[] complete = new String[0];
complete = creatures.toArray(complete);
for(String item : complete) {
    System.out.print(item + " ");
}
System.out.println();
```

When executed, we get the following output:

```
Mutant Godzilla Tribbles Ewoks Alien Zombie
```

There is also a `subList` method that returns part of the list given the starting and ending indexes.

Traversing a ArrayList object

To traverse a `ArrayList` object we can use one of several approaches:

- A simple for statement
- A for-each statement
- Using `Iterator`
- Using `ListIterator`

We can use a for loop but it is more prone to error. The following code will display the list from the beginning to the end:

```
for(int i = 0; i < creatures.size(); i++) {
    System.out.print(creatures.get(i) + " ");
}
System.out.println();
```

Notice the use of the `size` method, which returns the number of elements in the list.

The for-each statement is the simplest approach, as illustrated in the following code snippet:

```
for(String creature : creatures) {
    System.out.print(creature + " ");
}
System.out.println();
```

The `iterator` method returns a `Iterator` object, as shown below:

```
Iterator<String> iterator = creatures.iterator();
while(iterator.hasNext()) {
    System.out.print(iterator.next() + " ");
}
System.out.println();
```

The `ListIterator` method returns a `ListIterator` object:

```
ListIterator<String> listIterator =
            creatures.listIterator();
while(listIterator.hasNext()) {
    System.out.print(listIterator.next() + " ");
}
System.out.println();
```

All four of these techniques will produce the same output as follows:

`Mutant Godzilla Tribbles Ewoks Alien Zombie`

If we add the following code to the end of the previous code sequence, we can traverse the list in reverse order, as shown in the following code snippet:

```
while(listIterator.hasPrevious()) {
    System.out.print(listIterator.previous() + " ");
}
System.out.println();
```

The output is as follows:

`Zombie Alien Ewoks Tribbles Godzilla Mutant`

Sorting a ArrayList object

While there are no specific methods in the `ArrayList` class for sorting, we can use the `Arrays` class' `sort` method, as illustrated in the following code snippet:

```
Collections.sort(creatures);
System.out.println(creatures);
```

The output is as follows:

`[Alien, Ewoks, Godzilla, Mutant, Tribbles, Zombie]`

An overloaded version of this method takes a `Comparator` object. This object determines how comparisons are made.

Other ArrayList methods

We can modify an element of a list using the `set` method. This method takes an index of the element to replace, and the new value. For example, to replace the first element of the creatures list with the string `Ghoul` we can use the following code:

```
creatures.set(0,"Ghoul");
System.out.println(creatures);
```

The replacement is verified by the following output:

`[Ghoul, Godzilla, Tribbles, Ewoks, Alien, Zombie]`

We can remove all or some of the elements of a list. The `clear` method will remove all elements. The `remove` method removes a single element and the `removeAll` method removes all values in a given collection from the list. The following code sequence illustrates these methods. The `cuddles ArrayList` was defined in the *Adding elements* section:

```
System.out.println(creatures);
creatures.remove(0);
System.out.println(creatures);

creatures.remove("Alien");
System.out.println(creatures);

creatures.removeAll(cuddles);
System.out.println(creatures);

creatures.clear();
System.out.println(creatures);
```

The output of this sequence is as follows:

```
[Mutant, Godzilla, Tribbles, Ewoks, Alien, Zombie]
[Godzilla, Tribbles, Ewoks, Alien, Zombie]
[Godzilla, Tribbles, Ewoks, Zombie]
[Godzilla, Zombie]
[]
```

While `ArrayList` is a powerful class, arrays should still be used if:

- There is a known number of elements
- It has a small fixed upper bound
- Primitive data types are needed for efficiency
- No elements need to be inserted

Encapsulating collections

When using a collection within a class, hide the collection to prevent inadvertent modification of the collection. For example, if a class encapsulates an `ArrayList` of `Books`, then public methods should be provided to permit access to the collection. In the following example, a class called `Library` hides an `ArrayList` of `Book` objects:

```java
public class Library {

    private ArrayList<Book> books = new ArrayList<Book>();

    public Book getBook(int index) {
        return books.get(index);
    }

    public void addBook(Book book) {
        books.add(book);
    }

    public List getBooks() {
        return books;
    }
}
```

This is a good example of data encapsulation. However, be sure to not inadvertently expose private data. In the `getBook` method we returned a reference to the book. This reference allows the user to modify the book. If this modification should not be allowed, then a copy of the book can be returned instead, as shown below. This assumes that the `Book` class has a constructor that makes a new copy of a book based upon the constructor's argument:

```
public Book getBook (int index) {
    return new Book(books.get(index));
}
```

The same problem occurs with the `getBooks` method. It returns a reference to the private `books` reference variable of the `Library` class. This method can be replaced with the following implementation to ensure proper data encapsulation:

```
public List getBooks() {
    ArrayList list = new ArrayList(books.size());
    for(Book book : books) {
        list.add(new Book(book));
    }
    return list;
}
```

Summary

In this chapter we examined the creation and use of arrays and instances of the `ArrayList` class. We also detailed the use of the `Arrays` class in support of various array operations.

Arrays contain one or more dimensions and are treated as objects. Care must be taken while using arrays to avoid problems accessing their elements. Problems can be avoided with a good understanding of how arrays are allocated in memory and of how to perform various operations, such as copying and comparing arrays, on them. Arrays are useful when we need a list of a fixed size as it allows efficient access of its elements.

The `Arrays` class provides a number of static methods that support arrays. For example, we can use the `Arrays` class to make copies of arrays, sort arrays, and fill arrays.

The `ArrayList` class provides an alternate approach for dealing with lists of data. It provides numerous methods for manipulating a list and will grow as needed when new elements are added to the list. This is one of its primary advantages over arrays. As with most data structures, it is important to encapsulate information in a class to help reduce the complexity of software development.

Now that we've learned about arrays, we're ready to look more carefully at the various looping constructs available in Java. We will examine these constructs in the next chapter.

The Collections Framework introduced several new interfaces and classes to replace older versions in the `java.util` package. We examined the `ArrayList` class and its methods used to manipulate its elements. The `ArrayList` class is more flexible than an array and is particularly useful for inserting and removing elements.

Certification objectives covered

In this chapter we covered the following certification objectives:

- Using one-dimensional arrays
- Using multidimensional arrays
- Declaring and using `ArrayList`

Test your knowledge

1. Which of the following statements will compile without an error?

    a. `int arr[];`
    b. `int arr[5];`
    c. `int arr[5] = {1,2,3,4,5};`
    d. `int arr[] = {1,2,3,4,5};`

2. Which of the following declares an array that supports two rows and a variable number of columns?

    a. `int arr[][] = new int[2][3];`
    b. `int arr[][] = new int[2][];`
    c. `int arr[][] = new int[][];`
    d. `int arr[][] = new int[][3];`

3. Given the following code, which of the following statements can be used to determine if cat can be found in the list?

```
ArrayList<String> list = new ArrayList<>();
list.add("dog");
list.add("cat");
list.add("frog");
```

   a. `list.contains("cat")`

   b. `list.hasObject("cat")`

   c. `list.indexOf("cat")`

   d. `list.indexOf(1)`

5
Looping Constructs

It is often desirable to repeat a sequence of actions again and again. For example, we may want to display information about the employees in an organization stored in an array. Each element of the array might hold a reference to a `Employee` object. A call to methods of the object would be placed inside a looping construct.

In Java there are four looping constructs available:

- For statement
- For-each statement
- While statement
- Do while statement

In addition, the break and continue statements are used within a loop to control how the loop behaves. The break statement is used to prematurely exit or short circuit the loop and is discussed in the *The break statement* section. As we observed in the *The switch statement* section in *Chapter 3, Decision Constructs*, the break is also used within the switch statement. The continue statement is used to bypass statements in a loop and continue executing the loop. It is covered in the *The continue statement* section. We will also examine the use of labels in Java, though they should be used sparingly.

The body of the loop is iterated through a specific number of times based on the loop structure. Iteration is the term commonly used to describe this execution.

Loops use control information to determine how many times the body of the loop will be executed. For most loops there is an initial set of values, a set of operations to be performed at the end of the body, and a terminal condition which will stop the execution of the loop. Not all loops have all of these parts, as some of these parts are either missing or implied. The terminal condition is almost always present as this is needed to terminate the iteration of the loop. If the terminal condition is missing, an infinite loop is created.

Infinite loops refer to those loops that may never terminate without using a statement, such as the break statement. Despite their name, infinite loops do not execute indefinitely as they will always terminate at some point. They are useful in situations where it is inconvenient or awkward to provide a loop termination condition as a part of the basic loop construct.

We will also cover the use of nested loops and various pitfalls associated with loops. A section dealing with the development of programming logic is also presented to help provide an approach when creating the program logic.

The for statement

The for statement is used when the number of times the loop needs to be executed is known. There are two variations of the for loop. The first one is discussed in this section and is the traditional form. The for-each statement is the second form and was introduced in Java 5. It is discussed in the *The for-each statement* section.

The for statement consists of the following three parts:

- Initial operation
- Terminal condition
- End loop operation

The general form of the for loop follows:

```
for (<initial-expression>;<terminal-expression>;<end-loop
operation>)
   //statements;
```

The body of a for loop is typically a block statement. The initial operation takes place prior to the first iteration of the loop and is executed only once. The end loop operations take place at the end of each execution of the loop. The terminal condition determines when the loop will terminate and is a logical expression. It is executed at the beginning of each repetition of the loop. Thus, the body of the for loop may be executed zero times if the first time the terminal condition is evaluated, it evaluates to false.

A variable is normally used as part of the initial operation, terminal condition, and end loop operation. The variable is either declared as part of the loop or is declared external to the loop. The following code snippet is an example of declaring a variable, `i`, as part of the loop. An example of using an external variable is covered in the *The for statement and scope* section:

```
for (int i = 1; i <= 10; i++) {
   System.out.print(i + " ");
}
System.out.println();
```

In this example we used a single statement in the body of the loop. The variable `i` was assigned an initial value of 1 and is incremented by 1 each time the loop executes. The loop executed 10 times and produced 1 line of output. The statement, `i++`, is a more concise way of saying `i = i + 1`. The output should be the following:

```
1  2  3  4  5  6  7  8  9  10
```

The following example uses a for statement to compute the square of the integers from `1` to `64`:

```
for (int i = 1; i <= 64; i++) {
   System.out.println (i + " squared is = " + i * i);
}
```

A partial listing of the output follows:

```
1 squared is = 1
2 squared is = 4
3 squared is = 9
4 squared is = 16
...
```

The initial value of the loop variable can be any value. In addition, the end loop operation can decrement or otherwise modify the variable as needed. In the next example, numbers are displayed from `10` to `1`:

```
for (int i = 10; i > 0; i--) {
   System.out.print(i + " ");
}
System.out.println();
```

The output of this sequence follows:

```
10  9  8  7  6  5  4  3  2  1
```

Looping Constructs

A common operation is to compute a cumulative sum, as illustrated with the following code sequence. This example is discussed in more detail in the *Timing is everything* section:

```
int sum = 0;
for(i = 1; i <= 10; i++) {
   sum += i;
}
System.out.println(sum);
```

The value of `sum` should be `55`.

The comma operator

The comma operator can be used as part of a for statement to add other variables for use within the loop and/or to control the loop. It is used to separate the parts of the initial-expression and the end-loop operation sections of the for loop. The use of the comma operator is shown as follows:

```
for(int i = 0, j = 10; j > 5; i++, j--) {
    System.out.printf("%3d   %3d%n",i , j);
}
```

Notice the use of the `%n` format specifier in the `printf` statement. This specifies that a new line character should be generated. In addition, this new line separator is platform-specific making the application more portable.

When executed, this code sequence will produce the following output:

```
0    10
1     9
2     8
3     7
4     6
```

Two variables were declared for the loop, `i` and `j`. The variable `i` was initialed to `0` and `j` was initialized to `10`. At the end of the loop, `i` was incremented by `1` and `j` was decremented by `1`. The loop executed as long as `j` was greater than `5`.

We could have used a more complex terminal condition, such as illustrated in the following code snippet:

```
for(int i = 0, j = 10; j > 5 && i < 3; i++, j--) {
    System.out.printf("%3d   %3d%n",i , j);
}
```

In this example, the loop will terminate after the third iteration resulting in the following output:

```
0    10
1    9
2    8
```

It is illegal to declare the variables separately, as attempted here:

```
for(int i = 0, int j = 10; j > 5; i++, j--) {
```

A syntax error is generated, shown as follows. Only the first part of the message is provided as it is lengthy. This also illustrates the cryptic nature of error messages generated by Java and most other programming languages:

```
<identifier> expected
'.class' expected
...
```

The for statement and scope

The index variable used by a for statement can have different scope depending on how it is declared. We can use this to control the execution of the loop and then use the variable outside the loop, as needed. The first example of a for loop is repeated as follows. In this code sequence the scope of the i variable is restricted to the body of the for loop:

```
for (int i = 1; i <= 10; i++) {
    System.out.println(i);
}
System.out.println();
```

An alternate approach declares i external to the loop as follows:

```
int i;
for (i = 1; i <= 10; i++) {
   System.out.print(i + " ");
}
System.out.println();
```

These two for loops are equivalent as they both display the numbers 1 to 10 on a single line. They differ in the scope of the `i` variable. In the first example, the scope is restricted to the body of the loop. An attempt to use the variable outside of the loop, as illustrated in the following code, will result in a syntax error:

```
for (int i = 1; i <= 10; i++) {
   System.out.println(i);
}
System.out.println(i);
```

The error message follows:

cannot find symbol

symbol: variable i

In the second example, upon termination of the loop the variable will retain its value and will be available for subsequent use. The following example illustrates this:

```
int i;
for (i = 1; i <= 10; i++) {
   System.out.print(i + " ");
}
System.out.println();
System.out.println(i);
```

The output of this sequence follows:

```
1  2  3  4  5  6  7  8  9  10
11
```

Scope is discussed in more detail in the *Scope and lifetime* section in *Chapter 2, Java Data Types and Their Usage*.

The for loop variations

The for loop may have a body consisting of multiple statements. It is important to remember that the for loop body consists of a single statement. The following illustrates the use of multiple statements in a loop. This loop will read in a sequence of numbers and print them out one per line. It will continue until it reads in a negative value and then it will exit the loop. The `java.util.Scanner` class is used to read in data from the input source. In this case it uses `System.in` which specifies the keyboard as its input source:

```
Scanner scanner = new Scanner(System.in);
int number = 0;

for (int i = 0; number >= 0; i++) {
```

```
        System.out.print("Enter a number: ");
        number = scanner.nextInt();
        System.out.printf("%d%n", number);
    }
```

One possible output of executing this code sequence is as follows:

Enter a number: 3

3

Enter a number: 56

56

Enter a number: -5

-5

The initial operation, terminal condition, or end loop operation are not required. For example, the following statements will execute the i++ statement 5 times with a value 5 assigned to i upon the exit from the loop:

```
int i = 0;
for (;i<5;) {
    i++;
}
```

In the following example, the body of the loop will execute forever creating an infinite loop:

```
int i = 0;
for (;;i++)
    ;
```

The same is true for the following for loop:

```
int i = 0;
for(;;)
    ;
```

This is called an **infinite loop** and is covered in more detail in the *Infinite loops* section.

> The for loop is normally used when you know how many times the loop will be executed. A controlling integer variable is typically used as an index into an array or for computational purposes within the body of the loop.

Looping Constructs

The for-each statement

The for-each statement was introduced with the release of Java 5. It is sometimes referred to as the enhanced for loop. Advantages of using the for-each statement include:

- It is unnecessary to provide end conditions for the counter variable
- It is simpler and more readable
- The statement provides opportunities for compiler optimization
- The use of generics is simplified

The for-each statement is used in conjunction with collections and arrays. It provides an easier way to iterate through each member of an array or class that has implemented the `java.util.Iterable` interface. As the `Iterable` interface is the super interface of the `java.util.Collection` interface, the for-each statement can be used with those classes that implement the `Collection` interface.

The syntax of this statement is similar to the regular for statement, except for the contents of its parentheses. The contents include a data type followed by a variable, a colon, and then an array name or collection, illustrated as follows:

```
for (<dataType variable>:<collection/array>)
    //statements;
```

Its use with a collection is illustrated in the *Using the for-each statement with a list* section. In the following sequence, an array of integers is declared, initialized, and a for-each statement is used to display each element of the array:

```
int numbers[] = new int[10];
for (int i = 0; i < 10; i++) {
    numbers[i] = i;
}
for (int element : numbers) {
    System.out.print(element + " ");
}
System.out.println();
```

The elements of the numbers array were initialized to their index. Notice that a for statement was used. This was because we are unable to access an index variable directly in a for-each statement easily. The for-each statement in the preceding code snippet is read as **"for each element in numbers"**. During each iteration of the loop, `element` corresponds to an element of the array. It starts with the first element and ends with the last element. The output of this sequence is as follows:

```
0 1 2 3 4 5 6 7 8 9
```

There are drawbacks to the use of the for-each statement with an array. It is not possible to do the following:

- Modify the current position in an array or list
- Directly iterate over multiple arrays or collections

For example, using the previous example, if we try to modify the element of the array containing a 5 with the following code, it will not result in a syntax error. But it also will not modify the corresponding array element:

```
for (int element : numbers) {
   if (element == 5) {
      element = -5;
   }
}

for (int element : numbers) {
   System.out.print(element + " ");
}
System.out.println();
```

The output of this sequence is as follows:

```
0 1 2 3 4 5 6 7 8 9
```

If we want to use one loop to access two different arrays, the for-each loop cannot be used. For example, if we want to copy one array to another, we need to use the for loop, shown as follows:

```
int source[] = new int[5];
int destination[] = new int[5];

for(int number : source) {
   number = 100;
}

for(int i = 0; i < 5; i++) {
   destination[i] = source[i];
}
```

While we used a for-each to initialize the source array, we can only address a single array at a time. Thus, in the second loop we were forced to use the for statement.

Using the for-each statement with a list

We will start by illustrating the use of the for-each statement with the `ArrayList`. The `ArrayList` class implements the `List` interface which extends the `Collection` interface. The use and declaration of interfaces is addressed in more detail in *Chapter 6, Classes, Constructors, and Methods*. As the for-each statement can be used with classes that implement the `Collection` interface, we can also use it with the `ArrayList` class. In the next section, we will create our own `Iterable` class:

```
ArrayList<String> list = new ArrayList<String>();

list.add("Lions and");
list.add("tigers and");
list.add("bears.");
list.add("Oh My!");

for(String word : list) {
    System.out.print(word + " ");
}
System.out.println();
```

The output, as you might predict, is as follows:

Lions and tigers and bears. Oh My!

The use of the for-each in this example is not that much different from its use with an array. We simply used the name of the `ArrayList` instead of an array name.

Using a for-each statement with a list has similar restrictions to those we saw earlier with arrays:

- May not be able to remove elements from a list as you traverse it
- Inability to modify the current position in a list
- Not possible to iterate over multiple collections

The `remove` method can throw a `UnsupportedOperationException` exception. This is possible because the implementation of the `Iteratable` interface's `Iterator` may not have implemented the `remove` method. This is elaborated on in the next section.

In the case of the `ArrayList`, we can remove an element, as demonstrated in the following code snippet:

```
for(String word : list) {
    if(word.equals("bears.")) {
        list.remove(word);
        System.out.println(word + " removed");
```

```
        }
    }
    for(String word : list) {
        System.out.print(word + " ");
    }
    System.out.println();
```

The for-each statement was used to iterate over the list. When the `bears.` string was found, it was removed. The output of the preceding sequence is as follows:

```
Lions and tigers and bears. Oh My!
bears. removed
Lions and tigers and Oh My!
```

We cannot modify the list from within the for-each statement. For example, the following code sequence attempts to modify `word` and add a string to `list`. The list will not be affected:

```
    for(String word : list) {
        if(word.equals("bears.")) {
            word = "kitty cats";
            list.add("kitty cats");
        }
    }
```

While the attempt to modify the `word` variable does not do anything, it does not generate an exception. This is not the case with the `add` method. When used within the preceding for-each statement, it will generate a `java.util.ConcurrentModificationException` exception.

> As with arrays, it is not possible to iterate over more than one collection at a time using the for-each statement. As the for-each statement supports only one reference variable, only one list can be accessed at a time.
>
> If you need to remove an element from a list, use an iterator instead of a for-each statement.

Implementing the Iterator interface

As mentioned earlier, any class that implements the `Iterable` interface can be used with the for-each statement. To illustrate this we will create two classes:

- `MyIterator`: This implements the `Iterator` interface and supports a trivial iteration
- `MyIterable`: This uses `MyIterator` to support its use in a for-each statement

First, let's examine the `MyIterator` class that follows. The class will iterate through the numbers 1 to 10. It does this by comparing the `value` variable against an upper bound of 10 and returning either `true` or `false` in its `hasNext` method. The `next` method simply returns and increments the current value. The `remove` method is not supported:

```java
import java.util.Iterator;

public class MyIterator implements Iterator<Integer> {
    private int value;
    private final int size;

    public MyIterator() {
        value = 1;
        size = 10;
    }

    @Override
    public boolean hasNext() {
        return value<=size;
    }

    @Override
    public Integer next() {
        return value++;
    }

    @Override
    public void remove() {
        throw new UnsupportedOperationException(
            "Not supported yet.");
    }
}
```

The `MyIterable` class implements the `Iterable` interface. This interface consists of a single method, `iterator`. In this class, it uses an instance of the `MyIterator` class to provide a `Iterator` object:

```
import java.util.Iterator;

public class MyIterable implements Iterable<Integer> {
   private MyIterator iterator;

   public MyIterable() {
      iterator = new MyIterator();
   }

   @Override
   public Iterator<Integer> iterator() {
      return iterator;
   }
}
```

We can test these classes with the following code sequence:

```
MyIterable iterable = new MyIterable();

for(Integer number : iterable) {
   System.out.print(number + " ");
}
System.out.println();
```

The output will display the numbers from 1 to 10, shown as follows:

```
1 2 3 4 5 6 7 8 9 10
```

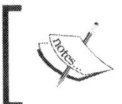

> The use of the `Iterator` methods for iterating through a collection is not always needed. In many situations, the for-each statement provides a much more convenient and simple technique.

The for-each statement – usage issues

There are several issues that you should be aware of when working with the for-each statement:

- If the array/collection is null, you will get a null pointer exception
- It works well with a method having a variable number of arguments

Null values

If the array/collection is null, you will get a null pointer exception. Consider the following example. We create an array of strings but fail to initialize the third element:

```
String names[] = new String[5];
names[0] = "Will Turner";
names[1] = "Captain Jack Sparrow";
names[3] = "Barbossa";
names[4] = "Elizabeth Swann";
```

We can display the names using a for-each statement as follows:

```
for(String name : names) {
    System.out.println(name);
}
```

The output, shown as follows, will display `null` for the missing entry. This is because the `println` method checks its argument for a null value and when it is, it prints `null`:

Will Turner

Captain Jack Sparrow

null

Barbossa

Elizabeth Swann

However, if we apply the `toString` method against the name as follows, we will get `java.lang.NullPointerException` on the third element:

```
for(String name : names) {
    System.out.println(name.toString());
}
```

This is verified, as shown in the following output:

Will Turner

Captain Jack Sparrow

java.lang.NullPointerException

Variable number of arguments

The for-each statement works well in methods using a variable number of arguments. A more detailed explanation of methods that use a variable number of arguments is found in the *Variable number of arguments* section in *Chapter 6, Classes, Constructors, and Methods*.

In the following method we pass a variable number of integer arguments. Next, we calculate the cumulative sum of these integers and return the sum:

```
public int total(int ... array) {
    int sum = 0;
    for(int number : array) {
        sum+=number;
    }
    return sum;
}
```

When this is executed with the following calls, we get `15` and `0` as output:

```
result = total(1,2,3,4,5);
result = total();
```

However, we need to be careful not to pass a `null` value as this will result in `java.lang.NullPointerException`, as illustrated in the following code snippet:

```
result = total(null);
```

> Use the for-each loop whenever possible, instead of the for loop.

The while statement

The while statement provides an alternate way of repeatedly executing a block of statements. It is frequently used when the number of times the block is to be executed is not known. Its general form consists of the `while` keyword followed by a set of parentheses enclosing a logical expression and then a statement. The body of the loop will execute as long as the logical expression evaluates to true:

```
while (<boolean-expression>) <statements>;
```

Looping Constructs

A simple example duplicates the first for loop example where we display the numbers 1 to 10 on a single line:

```
int i = 1;
while(i <= 10) {
    System.out.print(i++ + " ");
}
System.out.println();
```

The output is as follows:

```
1 2 3 4 5 6 7 8 9 10
```

The following example is a bit more complicated and computes the factors of the `number` variable:

```
int number;
int divisor = 1;
Scanner scanner = new Scanner(System.in);
System.out.print("Enter a number: ");
number = scanner.nextInt();
while (number >= divisor) {
    if ((number % divisor) == 0) {
        System.out.printf("%d%n", divisor);
    }
    divisor++;
}
```

When executed with the input as 6, we get the following output:

```
Enter a number: 6
1
2
3
6
```

The following table illustrates the action of the statement's sequence:

Iteration count	divisor	number	Output
1	1	6	1
2	2	6	2
3	3	6	3
4	4	6	
5	5	6	
6	6	6	6

In the following example, the loop will terminate when the user types in a negative number. In the process, it calculates the cumulative sum of the numbers entered:

```
int number;
System.out.print("Enter a number: ");
number = scanner.nextInt();
while (number > 0) {
   sum += number;
   System.out.print("Enter a number: ");
   number = scanner.nextInt();
}
System.out.println("The sum is " + sum);
```

Notice how this example duplicated the code needed to prompt the user for a number. The problem can be handled more elegantly using a do-while statement as discussed in the next section. The following output illustrates the execution of this code for a series of numbers:

```
Enter a number: 8
Enter a number: 12
Enter a number: 4
Enter a number: -5
The sum is 24
```

The while statement is useful for loops where the number of loop iterations required is not known. The body of the while loop will execute until the loop expression becomes false. It is also useful when the terminal condition is rather complex.

> An important characteristic of the while statement is the evaluation of the expression at the beginning of the loop. As a result, the body of the loop may never be executed if the first evaluation of the logical expression evaluates to false.

The do-while statement

The do-while statement is similar to a while loop except that the body of the loop always executes at least once. It consists of the `do` keyword followed by a statement, the `while` keyword, and then a logical expression enclosed in parentheses:

```
do <statement> while (<boolean-expression>);
```

Looping Constructs

Typically, the body of the do-while loop, as represented by the statement, is a block statement. The following code snippet illustrates the use of the do statement. It is an improvement over the equivalent while loop used in the previous section, as it avoids prompting for a number before the loop starts:

```
int sum = 0;
int number;
Scanner scanner = new Scanner(System.in);
do {
   System.out.print("Enter a number: ");
   number = scanner.nextInt();
   if(number > 0 ) {
      sum += number;
   }
} while (number > 0);
System.out.println("The sum is " + sum);
```

When executed you should get output similar to the following:

```
Enter a number: 8
Enter a number: 12
Enter a number: 4
Enter a number: -5
The sum is 24
```

> The do-while statement differs from that of the while statement as the evaluation of the expression occurs at the end of the loop. This means that this statement will be executed at least once.

This statement is not used as frequently as the for or while statement, but is useful in situations where a test at the bottom of a loop is best. The next statement sequence will determine the number of digits in an integer number:

```
int numOfDigits;
System.out.print("Enter a number: ");
Scanner scanner = new Scanner(System.in);
int number = scanner.nextInt();
numOfDigits = 0;
do {
   number /= 10;
   numOfDigits++;
} while (number != 0);
System.out.printf("Number of digits: %d%n", numOfDigits);
```

The output of this sequence follows:

```
Enter a number: 452
Number of digits: 3
```

The result for the value 452 is illustrated in the following table:

Iteration count	number	numOfDigits
0	452	0
1	45	1
2	4	2
3	0	3

The break statement

The effect of the break statement is to terminate the current loop, whether it be a while, for, for-each, or do-while statement. It is also used in the switch statement. The break statement passes control to the next statement following the loop. The break statement consists of the break keyword.

Consider the effect of the following statement sequence which repeatedly prompts the user for a command within an infinite loop. The loop will be terminated when the user enters the Quit command:

```
String command;
while (true) {
   System.out.print("Enter a command: ");
   Scanner scanner = new Scanner(System.in);
   command = scanner.next();
   if ("Add".equals(command)) {
      // Process Add command
   } else if ("Subtract".equals(command)) {
      // Process Subtract command
   } else if ("Quit".equals(command)) {
      break;
   } else {
      System.out.println("Invalid Command");
   }
}
```

Notice how the equals method is used. The equals method is executed against the string literal and the command is used as its argument. This approach avoids NullPointerException that will result if the command contains a null value. As the string literals are never null, this exception will never occur.

The continue statement

The continue statement is used to transfer control from inside a loop to the end of the loop but does not exit the loop like the break statement does. The continue statement consists of the keyword, `continue`.

When executed, it forces the evaluation of the loop's logical expression. In the following statement sequence:

```
while (i < j) {
    ...
    if (i < 0) {
        continue;
    }
    ...
}
```

if `i` is less than `0`, it will bypass the rest of the body of the loop. If the loop condition `i<j` does not evaluate to false, the next iteration of the loop will be performed.

The continue statement is often used to eliminate a level of nesting which is often necessary. The preceding example would appear as follows, if the continue statement was not used:

```
while (i < j) {
    ...
    if (i < 0) {
        // Do nothing
    } else {
        ...
    }
}
```

Nested loops

Loops can be nested within each other. Any nested combination of the for, for-each, while, or do-while loops is permitted. This is useful for addressing a number of problems. The example that follows computes the sum of the elements of a row in a two-dimensional array. It starts by initializing each element to the sum of its indexes. The array is then displayed. This is followed by nested loops to compute and display the sum of the elements for each row:

```
final int numberOfRows = 2;
final int numberOfColumns = 3;
int matrix[][] = new int[numberOfRows][numberOfColumns];
```

```java
for (int i = 0; i < matrix.length; i++) {
    for (int j = 0; j < matrix[i].length; j++) {
        matrix[i][j] = i + j;
    }
}

for (int i = 0; i < matrix.length; i++) {
    for(int element : matrix[i]) {
        System.out.print(element + " ");
    }
    System.out.println();
}

for (int i = 0; i < matrix.length; i++) {
    int sum = 0;
    for(int element : matrix[i]) {
        sum += element;
    }
    System.out.println("Sum of row " + i + " is " +sum);
}
```

Notice the use of the `length` method used to control the number of times the loops are executed. This makes the code more maintainable if the size of the arrays change. When executed we get the following output:

```
0 1 2
1 2 3
Sum of row 0 is 3
Sum of row 1 is 6
```

Notice the use of the for-each statement when the array is displayed and the sum of the rows are calculated. This simplifies the calculations.

The break and continue statements can also be used within nested loops. However, they will only be used in conjunction with the current loop. That is, a break out of an inner loop will only break out of the inner loop and not the outer loop. As we will see in the next section, we can break out of the outer loop from an inner loop using labels.

In the following modification of the last nested loop sequence, we break out of the inner loop when the sum exceeds 2:

```java
for (int i = 0; i < matrix.length; i++) {
    int sum = 0;
    for(int element : matrix[i]) {
        sum += element;
```

```
        if(sum > 2) {
            break;
        }
    }
    System.out.println("Sum of row " + i + " is " +sum);
}
```

The execution of this nested loop will change the sum of the last row as shown below:

```
Sum of row 0 is 3
Sum of row 1 is 3
```

The break statement took us out of the inner loop but not the outer loop. We can break out of the outer loop if there was a corresponding break statement within the immediate body of the outer loop. The continue statement behaves in a similar fashion in relation to inner and outer loops.

Using labels

Labels are names of locations within a program. They can be used to alter the flow of control and should be used sparingly. In the previous example, we were unable to break out of the inner most loop using the break statement. However, labels can be used to break us out of more than one loop.

In the following example, we place a label in front of the outer loop. In the inner loop, we execute the break statement when `i` is larger than 0 effectively terminating the outer loop after the sum has been calculated for the first row. A label consists of a name followed by a colon:

```
outerLoop:
for(int i = 0; i < 2; i++) {
    int sum = 0;
    for(int element : matrix[i]) {
        sum += element;
        if(i > 0) {
            break outerLoop;
        }
    }
    System.out.println("Sum of row " + i + " is " +sum);
}
```

The output of this sequence is as follows:

```
Sum of row 0 is 3
```

We can also use the continue statement with labels for a similar effect.

> Labels should be avoided as they can result in unreadable and hard to maintain code.

Infinite loops

An infinite loop is one that will execute forever unless a statement, such as the break statement is used to force its termination. Infinite loops are quite useful to avoid an awkward logical condition for a loop.

An infinite while loop should use the `true` keyword as its logical expression:

```
while (true) {
    // body
}
```

A for loop could be as simple as using nulls for each part of the for statement:

```
for (;;) {
    // body
}
```

A loop that never terminates would not normally be of value for most programs since most programs should eventually terminate. However, most infinite loops are designed to terminate using the break statement, shown as follows:

```
while (true) {
    // first part
    if(someCondition) {
        break;
    }
    // last part
}
```

Looping Constructs

This technique is fairly common and is used to simplify the logic of a program. Consider the need to read in an age and terminate when the age is negative. It is necessary to assign a non-negative value to age to ensure that the loop executes at least once:

```
int age;
age = 1;
Scanner scanner = new Scanner(System.in);
while (age > 0) {
    System.out.print("Enter an age: ");
    age = scanner.nextInt();
    // use the age
}
```

The other option is to duplicate the user prompt and the statement used to read in the age before the loop begins:

```
System.out.print("Enter an age: ");
age = scanner.nextInt();
while (age > 0) {
    System.out.print("Enter an age: ");
    age = scanner.nextInt();
    // use the age
}
```

Either an arbitrary value had to be assigned to age before the loop began, or it was necessary to duplicate code. Neither approach is satisfactory.

However, the use of an infinite loop results in cleaner code. No arbitrary value needs to be assigned and code does not need to be duplicated:

```
while (true) {
    System.out.print("Enter an age: ");
    age = scanner.nextInt();
    if (age < 0) {
        break;
    }
    // use the age
}
```

While there are many situations where an infinite loop is desirable, they can also occur when the programmer is not careful, resulting in unanticipated results. One common way is to build a for loop without a valid termination condition, illustrated as follows:

```
for(int i = 1; i > 0; i++) {
    // Body
}
```

The loop starts with a value of 1 for i and will increment i by 1 during each iteration of the loop. The termination condition suggests that the loop will not terminate as i will only get larger and would, thus, always be greater than 0. However, eventually the variable will overflow and i will become negative and the loop will terminate. How long this might take depends on the execution speed of the machine.

The moral of the story is, "Be careful with loops". Infinite loops can be both a useful construct for solving some problems and a problematic construct when used unintentionally.

Timing is everything

A common programming need is to perform some sort of summation. We have computed the sum of several sequences of numbers in previous examples. While the summation process is relatively straightforward, it can be difficult for novice programmers. More importantly, we will use it here to provide an insight to the programming process.

Programming, by its nature, is an abstract process. The programmer will need to look at a static code listing and infer its dynamic execution. This can be difficult for many people. One way of assisting the developer in writing code is to consider the following three issues:

- What do we want to do?
- How do we want to do it?
- When do we want to do it?

Here, we will ask and apply the answers to these three questions to address the summation problem. However, these questions are equally applicable to other programming problems.

Let's focus on calculating the average age for a group of students, which will involve the summation process. Assume the ages are stored in age array and then initialized as shown in the following code snippet:

```
final int size = 5;
int age[] = new int[size];
int total;
float average;
age[0] = 23;
age[1] = 18;
age[2] = 19;
age[3] = 18;
age[4] = 21;
```

Looping Constructs

Then the summation can be calculated as follows:

```
total = 0;
for (int number : age) {
    total = total + number;
}
average = total / (age.length * 1.0f);
```

Notice that `total` is explicitly assigned a zero value. Each iteration of the for loop will add the next `age` to `total`. At the completion of the loop, `total` will be divided by the length of the array times `1.0f` to compute the average. By using the array length the code expression does not need to be changed if the array size changes. Multiplying by `1.0f` is necessary to avoid integer division. The following table illustrates the variable's values as the loop executes:

Loop count	i	total
0	-	0
1	1	23
2	2	41
3	3	60
4	4	78
5	5	99

Let's examine this problem from the standpoint of the three basic questions, as follows:

- **What do we want to do**: We want to calculate the total pay for a department.
- **How do we want to do it**: This has a multipart answer. We know we need a variable to hold the total pay, `total`, and that it needs to be initialized to 0.

  ```
  total = 0;
  ```

  We also understand the basic operation to calculate the cumulative sum is as follows:

  ```
  total = total + number;
  ```

  The loop needs to use each element of the array so a for-each statement is used:

  ```
  for (int number : age) {
      ...
  }
  ```

We have the foundation for a solution to our problem.

- **When do we want to do it**: The "when" in this situation suggests three basic choices:
    - Before the loop
    - In the loop
    - After the loop

The three parts of our solution can be combined in different ways. The basic operation needs to be inside the loop because it needs to be performed more than once. Executing the basic operation only once will not result in an answer we will like.

The variable, total, needs to be initialized with a 0. How is this done? We do this by using an assignment statement. When should this be done? Before, in, or after the loop? Doing this after the loop would be silly. When the loop completes, total should contain the answer, not a zero. If we initialize it to 0 inside of the loop, then with each iteration of the loop, total is reset back to 0. That leaves us with placing the statement before the loop as the only option that makes sense. The first thing we want to do is to assign a 0 to total.

There seems to always be variations to the solutions of most problems. For example, we could have used a while loop instead of a for-each loop. The += operator could be used to shorten the basic operation. One potential solution that uses these techniques introduces an index variable:

```
int index = 0;
total = 0;

while(index < age.length) {
    total += age[index++];
}
average = total / (age.length * 1.0f);
```

Clearly, there is not always a best solution to a specific problem. This makes the programming process both a creative and potentially fun activity.

Pitfalls

As with most programming constructs, loops have their own set of potential pitfalls. In this section we will address areas that can present problems to the unwary developer.

Looping Constructs

One common problem occurs when programmers use a semicolon after every statement. For example, the following statement results in an infinite loop because of the extra semicolon:

```
int i = 1;
while(i < 10) ;
   i++;
```

The semicolon on a line by itself is the empty statement. This statement does nothing. However, in this example it constitutes the body of the while loop. The increment statement is not part of the while loop. It is the first statement that follows the while loop. Indention, while desirable, does not make the statement a part of the loop. Thus, `i` is never incremented and the logical control expression will always return true.

Failure to use a block statement for the body of a loop can be a problem. In the following example we attempt to calculate the sum of the product of the numbers from 1 to 5. However, this does not work properly because the body of the loop only encompasses the calculation of the product. The summation statement, when it is indented, is not part of the body of the loop and is only executed once:

```
int sum = 0;
int product = 0;
for(int i = 1; i <= 5; i++)
   product = i * i;;
   sum += product;
```

The correct implementation of the loop uses a block statement as shown below:

```
int sum = 0;
int product = 0;
for(int i = 1; i <= 5; i++) {
   product = i * i;;
   sum += product;
}
```

> It is always a good policy to use a block statement for the body of a loop, even if the body consists of a single statement.

In the following sequence the body of the loop consists of multiple statements. However, `i` is never incremented. This will also result in an infinite loop unless either limit is changed or a break statement is encountered:

```
int i = 0;
while(i<limit) {
  // Process i
}
```

[180]

Even simple-appearing loops may, in effect, be infinite loops if one is not careful with how floating-point arithmetic occurs. In this example, `0.1` is added to x with each iteration of the loop. The loop is supposed to stop when x exactly equals `1.1`. This will never occur because of issues in how floating point numbers are stored for certain values:

```
float x = 0.1f;
while (x != 1.1) {
    System.out.printf("x = %f%n", x);
    x = x + 0.1f;
}
```

The number `0.1` cannot be stored precisely in base two in the same way that the decimal equivalent of the fraction 1/3 cannot be represented exactly (0.333333…). The result of adding this number repeatedly to x will result in a number that is not quite `1.1`. The comparison, `x != 1.1`, will return true and the loop will never end. The output of the `printf` statement does not show this difference:

```
...
x = 0.900000
x = 1.000000
x = 1.100000
x = 1.200000
x = 1.300000
...
```

Be careful when working with operations that will involve auto-boxing. Depending on the implementation, it can result in a performance hit if boxing and un-boxing occurs frequently.

While not necessarily a pitfall, remember that logical expressions can short circuit. That is, the last part of a logical AND or OR operation may not be evaluated depending on the value returned from the evaluation of the first part. This is discussed in detail in the *Short circuit evaluation* section in *Chapter 3, Decision Constructs*.

> Remember that arrays, strings, and most collections are zero based. Forgetting to start the loop at 0 will overlook this first element.
>
> Always use a block statement as the body of a loop.

Summary

In this chapter we examined the support Java provides for loops. We have illustrated the use of the for, for-each, while, and do-while statements. These demonstrations provided insight into their correct usage, when they should be used, and when they should not be used.

The use of the break and continue statements was shown, along with the use of labels. We saw the utility of the break statement, in particular, in support of infinite loops. Labels, while they should be avoided, can be useful in breaking out of deeply nested loops.

Various pitfalls were examined and the creation of the summation process was studied to gain insight into general programming problems. Specifically, it addressed the question of where a code segment should be placed.

Now that we've learned about loops, we're ready to examine the creation of classes, methods, and data encapsulation, which is the topic of the next chapter.

Certification objectives covered

In this chapter we addressed the following certification objectives:

- Creating and using the while loops
- Creating and using the for loops including the enhanced for loop
- Creating and using the do/while loops
- Comparing the loop constructs
- Using break and continue

In addition, we provided additional coverage of these objectives:

- Define the scope of variables
- Use operators and decision constructs
- Declaring and using an ArrayList

Test your knowledge

1. Given the following declarations, which of the following statement will compile?

   ```
   int i = 5;
   int j = 10;
   ```

   a. `while(i < j) {}`
   b. `while(i) {}`
   c. `while(i = 5) {}`
   d. `while((i = 12)!=5) {}`

2. Given the following declaration of an array, which statement will display each element of the array?

   ```
   int arr[] = {1,2,3,4,5};
   ```

   a. `for(int n : arr[]) { System.out.println(n);   }`
   b. `for(int n : arr) { System.out.println(n);   }`
   c. `for(int n=1; n < 6; n++) {   System.out.println(arr[n]);  }`
   d. `for(int n=1; n <= 5; n++) {   System.out.println(arr[n]);  }`

3. Which of the following do/while loops will compile without errors?

   a.
   ```
   int i = 0;
   do {
       System.out.println(i++);
   } while (i < 5);
   ```

   b.
   ```
   int i = 0;
   do
       System.out.println(i++);
   while (i < 5);
   ```

   c.
   ```
   int i = 0;
   do
       System.out.println(i++);
   while i < 5;
   ```

   d.
   ```
   i = 0;
   do
       System.out.println(i);
       i++;
   while (i < 5);
   ```

Looping Constructs

4. Which of the following loops are equivalent?

   a.
   ```
   for(String n : list) {
       System.out.println(n);
   }
   ```

   b.
   ```
   for(int n = 0; n < list.size(); n++ ){
       System.out.println(list.get(n));
   }
   ```

   c.
   ```
   Iterator it = list.iterator();
   while(it.hasNext()) {
       System.out.println(it.next());
   }
   ```

5. What will be output by the following code?

   ```
   int i;
   int j;
   for (i=1; i < 4; i++) {
      for (j=2; j < 4; j++) {
         if (j == 3) {
            continue;
         }
         System.out.println("i: " + i + " j: " + j);
      }
   }
   ```

   a.
   ```
   i: 1 j: 2
   i: 2 j: 2
   i: 3 j: 2
   ```

   b.
   ```
   i: 1 j: 3
   i: 2 j: 3
   i: 3 j: 3
   ```

   c.
   ```
   i: 1 j: 1
   i: 2 j: 1
   i: 3 j: 1
   ```

6
Classes, Constructors, and Methods

In the heart of object-oriented programming are classes and the objects created from classes. The initialization of the objects occurs in constructors while the modification of the state of an object is carried through methods. The packaging of these constructors and methods is the focus of data encapsulation. The fundamentals of classes, constructors, methods, and data encapsulation are addressed in this chapter.

We start with an introduction to classes including a discussion of how objects are managed in memory. Common aspects of constructors and methods are then presented including the concept of a signature, the passing of arguments, and the uses of the `this` keyword.

The usage of constructors is discussed including default constructors, how they are overloaded, and the use of private constructors. The Java initialization sequence is covered including the use of initializer lists.

Methods and how they are used is explained including how to overload them and the creation of accessor and mutator methods. The chapter concludes with a discussion of static and instance methods.

Classes

A **class** is the definition of a data structure plus actions that operate on them which typically corresponds to a real world object or concept. A class is defined once but is not used directly within an application. Instead, objects are created (instantiated) based on a class and are allocated memory.

Throughout the chapter we will illustrate the use of constructors and methods using the `Employee` class. A part of this class is shown as follows:

```
public class Employee {
    private String name;
    private int zip;
    private int age;
    ...
}
```

This definition will be expanded to explain the concepts and techniques associated with classes and objects.

Object creation

Objects are created using the `new` keyword. The keyword is used in conjunction with a classname and results in memory being allocated from the heap for the object. The heap is a region of memory normally located "above" the stack as discussed in the *Stack and heap* section in *Chapter 2, Java Data Types and Their Usage*.

When a new object is instantiated using the `new` keyword:

- Memory is allocated for the new instance of the class
- A constructor is then called to perform initialization of the object
- A reference to the object is returned

In the following example, two instances of the `Employee` class are created and references are assigned to the reference variables, `employee1` and `employee2`:

```
Employee employee1 = new Employee();
Employee employee2 = new Employee();
```

Each instance of a class has its own set of instance variables that are independent of each other. This is shown in the following diagram. Notice that both instances of the class contain their own copies of the instance variables:

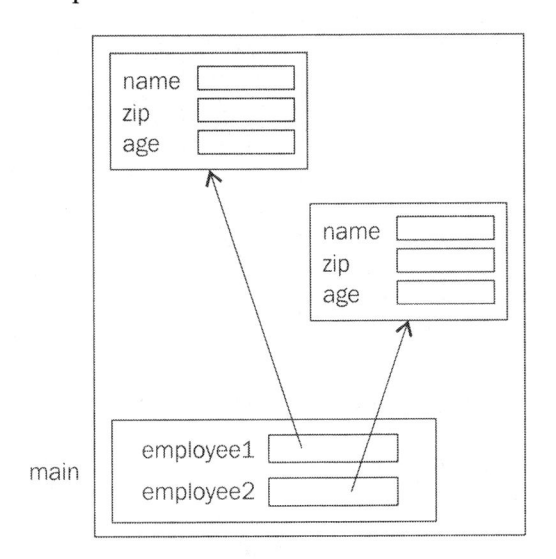

When a new object is created, a constructor for that object is executed. The purpose of a constructor is to initialize an object. This process is covered in the *Constructors* section. The class' methods are shared among the instances of the class. That is, there is only one copy of the methods.

Memory management

Java memory management is dynamic and automatic. When the new keyword is used, it automatically allocates memory on the heap.

In the following example, an instance of the Employee class is created and assigned to the employee1 variable. Next, the employee2 variable is assigned the value of the employee1 variable. The effect of this assignment is that both reference variables point to the same object:

```
Employee employee1 = new Employee();
Employee employee2 = employee1;
```

This is illustrated in the following diagram:

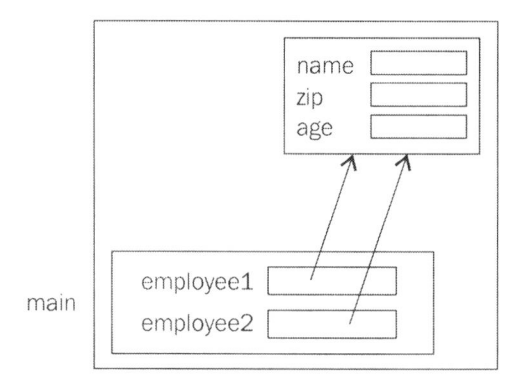

A reference variable may de-reference an instance of an object by:

- Being re-assigned to another object
- Setting it to null

When the garbage collector determines that there are no references to it, the object becomes eligible for removal from the heap by a garbage collection thread and its memory can be re-used for other objects. This garbage collection process is essentially beyond the control of the application.

Data encapsulation

Data encapsulation is concerned with hiding irrelevant information from the programmer and exposing the relevant information. Hiding the implementation details allow changes without affecting other parts of the program. For example, if a programmer wants to display a rectangle on the screen there are several approaches that can be used. It may involve drawing the rectangle pixel by pixel or drawing a series of lines. Hiding the details of the operation is referred to as data encapsulation.

The primary purpose of data encapsulation is to reduce the level of software development complexity. By hiding the details of what is needed to perform an operation, the use of that operation is simpler. The method is not that complex to use, as the user does not have to worry about the details of its implementation. The user can focus on what it does, not on how it does it. This, in turn, allows developers to do more.

For example, consider the implementation of the `Employee` class. Originally, the instance variables were both declared as private:

```
public class Employee {
    public String name;
    private int age;

    ...

    public int getAge() {
        return age;
    }

    private void setAge(int age) {
        this.age = age;
    }

}
```

The access modifier type of the `name` variable has been changed to public and the access modifier for the `setAge` method has been made private. This means that any user of the class can access the `name` field but they can only read the `age` of the employee. Data encapsulation has been affected when we explicitly decide what should and should not be exposed to the users of a class.

The details of a class and its implementation should be hidden from the user. This allows modification of the implementation of the class' internals without changing the public aspects of the class. As a general rule, instance variables are made private and methods are made public. Exceptions to this rule are made based on the needs of the class.

It is also possible to control access to constructors. This topic is covered in the *Constructors* section.

Referencing instance variables

A reference variable holds a reference, or pointer, to an object. A field or variable of the object is accessed by following the object reference variable name with a period and then the field or method name. The following code snippet illustrates possible references using the `Employee` class based upon the declaration of `Employee` found in the previous section:

```
Employee employee = new Employee();
int employeeAge = employee.getAge(24);
String employeeName = employee.name;
```

Classes, Constructors, and Methods

Notice that we did not use the `age` field as this was declared as private to the `Employee` class. The use of modifiers is covered in the *Access modifiers* section in *Chapter 1, Getting Started with Java*.

Signature

The signature of a constructor or method is used to uniquely identify a constructor or method. A signature consists of:

- Method or constructor name
- Number of parameters
- Type of the parameters
- Order of the parameters

All constructors or methods within the same class must have unique signatures. Note that the return type of the method is not part of a signature. The following table shows the signatures that overload the `Employee` class constructor. The third and fourth constructors differ in the order of the constructor's parameters. A method or constructor is said to be overloaded if there is more than one method or constructor with the same name, and in the same class, but with different signatures:

Method	Number of Arguments	Argument Types
`Employee()`	0	
`Employee(String name)`	1	`String`
`Employee(String name, int zip)`	2	`String, int`
`Employee(int zip, String name)`	2	`int, String`
`Employee(String name, int zip, int age)`	3	`String, int, int`

Using the this keyword

There are four uses of the `this` keyword:

- Performing constructor chaining
- Accessing instance variables
- Passing the current object to a method
- Returning the current object from a method

Constructor chaining is covered in the *Overloading constructors* section. Let's examine the use of the `this` keyword to access instance variables. The `setAge` method could have been implemented as follows:

```
public class Employee {
    public String name;
    private int age;
    ...

    private void setAge(int age) {
        age = age;
    }

}
```

This code would not have the intended consequences of modifying the `age` instance variable. The scope of the instance variables is the entire class. The scope of the parameters is only the method. The parameters will have "precedence" over the instance variables. The effect is that the age passed to the method was assigned to itself. The instance variable was not modified.

There are two ways of correcting this problem:

- Change the parameter name
- Use the `this` keyword

We could change the name of the parameter. However, devising a different name to designate the same thing leads to strange or awkward names. For example, we could have used the following method instead:

```
public class Employee {
  private int age;
      ...
    private void setAge(int initialAge) {
        age = initialAge;
    }

}
```

The `initialAge` parameter will be assigned as the initial value to the member variable `age`. However, any number of other potentially meaningful names could be used. There is no standard naming convention for naming parameters of this type.

Another approach is to declare the parameter as a constant using the `final` keyword, as shown in the following code snippet. When we take this approach, a syntax error is generated because we are trying to modify the parameter. As it is constant we cannot change it:

```
public void setAge(final int age) {
    age = age;
}
```

The syntax error message that is generated is as follows:

```
final parameter age may not be assigned
```

Assignment To Itself

The preferred approach is to use the `this` keyword to clearly specify which variable is the member variable and which is the parameter. This is illustrated in the following implementation:

```
public class Employee {
  private int age;
  ...
  private void setAge(int age) {
      this.age = age;
  }

}
```

In this assignment statement we referenced the member variable by prefixing it with the `this` keyword and a period. Consider the following statement:

```
    this.age = age;
```

The `this` keyword references the `age` instance variable, on the left-hand side of the assignment statement. On the right-hand side, the `age` parameter, was used. Thus, the parameter is assigned to the instance variable. Using the `this` keyword avoids having to devise some non-standard and potentially confusing name for parameters being assigned to a member variable.

The `this` keyword can also be used to pass or return a reference to the current object. In the following sequence, the `validateEmployee` method is assumed to be a member of the `Employee` class. If a condition is met, then the current employee, as identified by the `this` keyword, is added to a class maintaining department information as referenced by the `department` variable. A reference to the current object is passed to the `add` method:

```
private Department department;
    ...
private void validateEmployee() {
    if(someCondition) {
        department.add(this);
    }
}
```

The `this` keyword can also be used to return a reference to the current object. In the next sequence, the current object is returned by the `getReference` method which is assumed to be a method of the `Employee` class:

```
private Employee getReference() {
    ...
    return this;
}
```

Passing parameters

Within any method there may exist two types of variables—parameters and local variables. Parameters contain values passed to the method when it is invoked. Local variables are part of the method and are used to assist the method in the completion of its task. The techniques discussed here apply to both constructors and methods though we will only use methods for the examples in this section.

Parameters are passed as part of a parameter list. This list uses a comma to delimit the declaration of the type and name of a parameter. For example, the method in the following code snippet is passed two parameters—an integer and a string:

```
public void passParameters(int number, String label) {
    ...
}
```

Either a primitive data type or an object is passed to a method. The following terms are used to identify the data being passed:

- Argument: This is the variable being passed
- Parameter: This is the element defined within the method's signature

For example, in the following code sequence `number` and `employee1` are the arguments while `num` and `employee` are the corresponding parameters to the `changeValues` method:

```
public static void main(String[] args) {
    int number = 10;
    Employee employee1 = new Employee();
    changeValues(number, employee1);
```

```java
    ...
}

private static void changeValues(int num,
    Employee employee) {
    ...
}
```

In Java, only primitive data types and object references are passed to a method or constructor. This is performed using a technique called **passing by value**. When a method is called, a copy of the argument is assigned to the parameter.

When a primitive data type is passed, only a copy of the value is passed. This means if the copy in the called method is changed, the original data is not changed.

When a reference variable is passed, only a copy of the reference is passed. The object itself is not passed or copied. At this point we have two references to the same object—the argument reference variable and the parameter reference variable. We can modify the object using either reference variable.

We can also change what the parameter references. That is, we can modify the parameter to reference a different object. If we modify the parameter we are not modifying the argument. The parameter and the argument reference variables are distinct variables.

Consider the following program where we pass an integer and a reference to a `Employee` object, to the `changeValues` method. In the method we change the integer, a field of the `Employee` object, and the `employee` reference variable.

```java
public static void main(String[] args) {
    ...
    int number = 10;
    employee = new Employee();
    employee.setAge(11);
    changeValues(number, employee);

    System.out.println(number);
    System.out.println(employee.getAge());

}

private static void changeValues(int num,
        Employee employee) {
    num = 20;
    employee.setAge(22);
    employee = new Employee();
    employee.setAge(33);
}
```

When executed we get the following output:

```
10
22
```

> Notice that when we changed the value of the num parameter, the main method's number variable did not change. Also, we changed the object's age field using the changeValues method's employee reference variable. However, when we modified what the changeValues method's employee reference variable pointed to by creating a new employee, we did not change the main method's employee reference variable. It still references the original object.

The following diagram illustrates how this works. The stack and heap reflect the state of the application when the changeValues method is started and immediately before it returns. For simplicity, we have ignored the args variable:

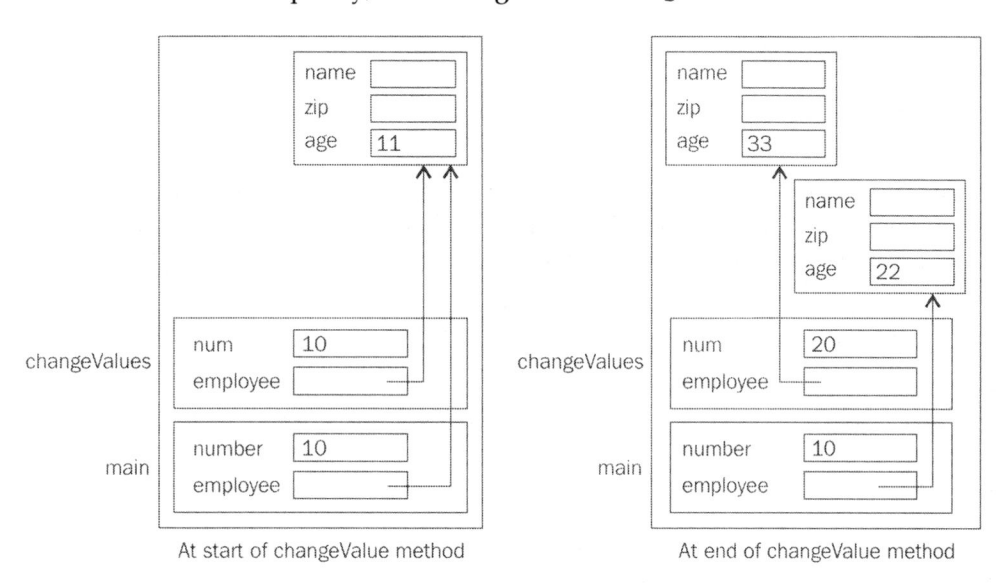

```
                    name
                    zip                              name
                    age   11                         zip
                                                     age   33
                                                              name
                                                              zip
                                                              age   22
changeValues  num     10         changeValues  num     20
              employee                         employee

main          number  10         main          number  10
              employee                         employee

   At start of changeValue method      At end of changeValue method
```

Passing an object by value is an efficient parameter passing technique. It is efficient because we are are not copying the entire object. We only copy the reference to the object.

Variable number of arguments

It is possible to pass a variable number of arguments to a method. However, there are some restrictions:

- The variable number of parameters must all be the same type
- They are treated as an array within the method
- They must be the last parameter of the method

To understand these restrictions, consider the method, in the following code snippet, used to return the largest integer in a list of integers:

```
private static int largest(int... numbers) {
    int currentLargest = numbers[0];
    for (int number : numbers) {
        if (number > currentLargest) {
            currentLargest = number;
        }
    }
    return currentLargest;
}
```

It is not necessary to declare methods with a variable number of parameters as static. We do this here so that we can call it from the static `main` method. In the following code sequence we invoke the method twice:

```
System.out.println(largest(12, -12, 45, 4, 345, 23, 49));
System.out.println(largest(-43, -12, -705, -48, -3));
```

The output is as follows:

345

-3

The `largest` method assigns the first parameter, the first element of the `numbers` array, to `currentLargest`. It makes the assumption that the largest number is the first parameter. If it is not, then it will eventually be replaced. This avoids having to assign the smallest possible value to the `currentLargest` variable.

> The largest and smallest integers are defined in the `Integer` class as `Integer.MAX_VALUE` and `Integer.MIN_VALUE` respectively.

We used a for-each statement to compare each element of the numbers array to the largest variable. If the number is larger, then we replace `largest` with that number. The for-each statement is detailed in the *The for-each statement* section of *Chapter 5, Looping Constructs*.

If we call the method with no arguments, as attempted below:

```
System.out.println(largest());
```

The program will execute but a `ArrayIndexOutOfBoundsException` exception will be generated. This occurs because we tried to access the first element of the array in the method which does not exist because the array is empty. If we had not referenced the first element in the method, this problem would not have occurred. That is, a method that uses a variable number of arguments can, in most circumstances, be called with no arguments.

We could have implemented a version of the `largest` method that handles the situation where no arguments are passed. However, when nothing is passed what should the return value be? Any value we returned would imply that that number is the largest when, in fact, there is not a largest number. The best we can probably do is to return an exception reflecting this problem. However, this is effectively what the current version does. The exception, `ArrayIndexOutOfBoundsException`, is perhaps not as meaningful as a custom exception.

We can use other parameters in a method possessing a variable number of arguments. In the following example we pass a string, and zero or more floats, to a `displayAspects` method. The intent of the method is to display information about the element identified by the first argument:

```
private static void displayAspects(String item,
    float... aspects) {
    ...
}
```

The following code is an example of how the method might be invoked:

```
displayAspects("Europa", 2.3f, 56.005f, 0.0034f);
```

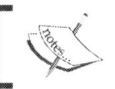

> Variable arguments must be all of the same type and must be the last ones in the parameter list.

Classes, Constructors, and Methods

Immutable objects

Immutable objects are objects whose state cannot be changed. By state, we mean the value of its member variables. These types of objects can simplify an application and are less error prone. There are several classes in the JDK core that are immutable including the `String` class.

To create an immutable object:

- Make the class final which means that it cannot be extended (covered in the *Using the final keyword with classes* section in *Chapter 7, Inheritance and Polymorphism*)
- Keep the fields of the class private and ideally final
- Do not provide any methods that modify the state of the object, that is do not provide setter or similar methods
- Do not allow mutable field objects to be changed

The following is an example of the declaration of an immutable class representing a header for a page:

```
package packt;

import java.util.Date;

final public class Header {
    private final String title;
    private final int version;
    private final Date date;

    public Date getDate() {
        return new Date(date.getTime());
    }

    public String getTitle() {
        return title;
    }

    public int getVersion() {
        return version;
    }

    public Header(String title, int version, Date date) {
        this.title = title;
        this.version = version;
```

```
            this.date = new Date(date.getTime());
    }

    public String toString() {
        return  "Title: " + this.title + "\n" +
                "Version: " + this.version + "\n" +
                "Date: " + this.date + "\n";
    }
}
```

Notice that the `getDate` method created a new `Date` object based on the header's `date` field. Any `Date` object is mutable, so by returning a copy of the date as opposed to a reference to the current date, the user is unable to access and otherwise modify the private field. The same approach was used in the three-argument constructor.

Constructors

Constructors are used to initialize the member variables of a class. When an object is created, memory is allocated for the object and the constructor for the class is executed. This typically occurs using the `new` keyword.

Initialization of an object's instance variables is important. One of the developer's responsibilities is making sure that the state of an object is always valid. To assist in this process, constructors are executed whenever an object is created.

An alternate approach, which is not used by Java, is to use an initialization method that the programmer should call after an object is created. However, the use of such an initialization method is not a foolproof technique. The programmer may not be aware that the method exists, or may forget to call the method. To avoid these types of problems, a constructor is automatically invoked when an object is created.

The important characteristics of constructors include:

- Constructors have the same name as the class
- Constructor overloading is permitted
- Constructors are not methods
- Constructors do not have a return type, not even void

The following code snippet illustrates how constructors are defined. In this example, three overloaded constructors are defined. For the moment, we have left out their bodies. The intent of these constructors is to initialize the three instance variables that make up the class:

```java
public class Employee {
    private String name;
    private int zip;
    private int age;

    public Employee() {

    }

    public Employee(String name) {

    }

    public Employee(String name, int zip) {

    }

}
```

Default constructors

A default constructor is normally present for a class. If a class does not have any constructors explicitly declared, it automatically has a default constructor. A default constructor is a constructor that has no arguments. This is illustrated in the following code snippet, for the `Employee` class where no constructors are defined:

```java
public class Employee {
    private String name;
    private int zip;
    private int age;

    ...

}
```

Chapter 6

The default constructor will essentially initialize its instance variables to 0 as explained in the *Initializing identifiers* section in *Chapter 2, Java Data Types and Their Usage*. The values assigned to member variables are found in the following table which is duplicated from *Chapter 2, Java Data Types and Their Usage*, for your convenience:

Data Type	Default Value (for fields)
boolean	false
byte	0
char	'\u0000'
short	0
int	0
long	0L
float	0.0f
double	0.0d
String (or any object)	null

However, we can also add an explicit default constructor, as shown in the following code snippet. The default constructor is a constructor that has no arguments. As we can see, we are free to initialize the fields of the class to whatever values we choose. For those fields that we do not initialize, the JVM will initialize them to zeroes as detailed above:

```
public Employee() {
    this.name = "Default name";
    this.zip = 12345;
    this.age = 21;
}
```

Note the use of the `this` keyword. In this context it is used to unambiguously specify that the variables following the period are class member variables, and not some other local variables. Here, there are no other variables that might cause confusion. The `this` keyword was detailed in the *Using the this keyword* section. It is a common practice to use the `this` keyword with member variables.

If the programmer adds a constructor to the class, then the class will no longer have a default constructor added automatically. The programmer must explicitly add a default constructor for the class to have one. In the following declaration of the `Employee` class, the default constructor has been left out:

```
public class Employee {
    private String name;
    private int zip;
    private int age;
```

```
        public Employee(String name) {
        }
        ...
    }
```

If we try to create an object using the default constructor, as shown in the following code snippet, then we will get a syntax error:

```
Employee employee1 = new Employee();
```

The error message that is generated is as follows:

no suitable constructor found for Employee()

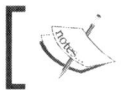

> As a general rule, always add a default constructor to a class. This is particularly important when the class is a base class.

Overloading the constructors

Constructors can be overloaded. By overloading the constructors, we provide the users of the class with more flexibility in how an object can be created. This can simplify the development process.

Overloaded constructors have the same name but different signatures. The definition of a signature was provided in the *Signature* section, discussed earlier. In the following version of the Employee class we provide four constructors. Notice how each constructor assigns default values for those member variables not passed with the constructor:

```
public class Employee {
    private String name;
    private int zip;
    private int age;

    public Employee() {
        this.name = "Default name";
        this.zip = 12345;
        this.age = 21;
    }
    public Employee(String name) {
        this.name = name;
        this.zip = 12345;
        this.age = 21;
    }
```

```java
    public Employee(String name, int zip) {
        this.name = name;
        this.zip = zip;
        this.age = 21;
    }


    public Employee(String name, int zip, int age) {
        this.name = name;
        this.zip = zip;
        this.age = age;
    }

}
```

This example duplicates work between the constructors. An alternate approach, shown as follows, uses the `this` keyword to reduce this duplication of effort and simplify the overall process:

```java
public class Employee {
    private String name;
    private int zip;
    private int age;

    public Employee() {
        this("Default name", 12345, 21);
    }

    public Employee(String name) {
        this(name, 12345, 21);
    }

    public Employee(String name, int zip) {
        this(name, zip, 21);
    }


    public Employee(String name, int zip, int age) {
        this.name = name;
        this.zip = zip;
        this.age = age;
    }
}
```

In this case, the `this` keyword is used at the beginning of a constructor with an argument list. The effect is to call the same class' constructor that matches the signature used. In this example, each of the first three constructors calls the last constructor. This is called **constructor chaining**. All of the work is performed in the last constructor reducing the amount of repeated work being performed and chances for errors, especially when new fields are added.

This can be even more productive if the field variables are checked within a constructor prior to their assignment. For example, if we need to verify that the name meets a specific set of naming criteria, it only needs to be performed in one location instead of each constructor that is passed a name.

Private constructors

A constructor can be declared as private which serves to hide it from the user. This may be done to:

- Restrict access to some, but not all, of the class' constructors
- Hide all of the constructors from a user

In some situations, we may desire to make constructors private or protected (See *Chapter 7, Inheritance and Polymorphism*, for a discussion of the `protected` keyword) to limit access to certain initialization sequences. For example, a private constructor may be used to initialize fields of a class in a less rigorous manner. As we are invoking the constructor from other constructors, we may be more confident of the values being assigned and do not feel that extensive checking of its parameters is needed.

It is not uncommon to find classes where all of the constructors are declared as private. This restricts the creation of objects by a user to the public methods of the class. The `java.util.Calendar` class is an example of such a class. The only way to obtain an instance of this class is to use its static `getInstance` method.

The use of private constructors is used to control the number of instances of the class that can be created by an application. The singleton design pattern dictates that only one instance of a class is ever created. This design pattern can be supported by making all of its constructors private and providing a public `getInstance` method that creates a single instance of the class.

The following illustrates this approach for the `Employee` class. The constructor is made private and the `getInstance` method insures that only a single object is ever created:

```
public class Employee {
   private static Employee instance = null;
   private String name;
   private int zip;
   private int age;

   private Employee instance = null;
   ...

   private Employee() {
      this.name = "Default name";
      this.zip = 12345;
      this.age = 21;
   }

   public Employee getInstance() {
      if(instance == null) {
         instance = new Employee();
      }
      return instance;
   }

   ...
}
```

The first time the `getInstance` method is called the `instance` variable is null, which results in a new `Employee` object being created. In subsequent calls to the `getInstance` method, `instance` will not be null and a new `Employee` object is not created. The current reference to the single object is returned.

Constructor issues

If a "constructor" has a return type, it is actually a method that happens to have the same name as the class. This is true even if the return type is `void`, as illustrated in the following code snippet:

```
public void Employee(String name) {

}
```

We can create a new instance of the `Employee` class and then apply the `Employee` method against this object, as shown in the following code snippet:

```
Employee employee = new Employee();
employee.Employee("Calling a method");
```

While this is legal, it is not good style and can be confusing. In addition, as we saw in the *Java naming conventions* section in *Chapter 1, Getting Started with Java*, the naming conventions for methods suggest that the initial word of a method's name should begin with a lowercase letter.

Java initialization sequence

Constructors are concerned with the initialization of the fields of an object. However, there are two other approaches that can be used to complement the use of constructors. The first is to use instance variable initializers. Using the `Employee` class we can initialize the age to 21, shown as follows:

```
public class Employee {
   ...
   private int age = 21;
   ...
}
```

If we initialize an instance variable in this manner, we do not have to initialize it in a constructor.

The second approach is to use an initialization block. This type of block is executed before the constructor is executed. The following code snippet illustrates this approach:

```
public class Employee {
   ...
   private int age;

   // Initialization block
   {
      age = 31;
   }
   ...
}
```

Initialization blocks are useful when more complex initialization sequences are needed which cannot be supported with the simpler instance variable initializers. This initialization can also be performed in a constructor.

Thus, there are several ways of initializing member variables. If we use one or more of these techniques to initialize the same variable, then we may wonder in what order they are performed. The actual initialization sequence is a bit more complex than described here. However, the general order is as follows:

1. The zeroing of fields performed when the object is instantiated
2. The initialization of final and static variables
3. The assignment of instance variables initializers
4. The execution of initialization blocks
5. The code within a constructor

More detail about the initialization sequence can be found in the Java Language Specification (http://docs.oracle.com/javase/specs/).

Methods

A **method** is a group of statements used to complete a specific task. A method has a return value, a name, a set of parameters, and a body. Parameters are passed to a method and are used to perform an action. If a value is to be returned from a method, the return statement is used. A method may have zero or more return statements. A method that returns void may use a return statement but the statement does not have an argument.

Defining methods

Methods are defined as part of the class definition and normally follow the declaration of the instance variables. The method declaration specifies a return type. The return type `void` means that the method does not return a value.

> The Java naming convention for methods specifies that the first word is not capitalized but subsequent words are capitalized. Method names should be verbs.

In the following example, the method returns `boolean` and is passed two integer parameters:

```
public boolean isAgeInRange(int startAge, int endAge) {
    return (startAge <= age) && (age <= endAge);
}
```

[207]

All methods within the same program must have unique signatures. Signatures are discussed in the *Signature* section, discussed earlier. Note that the return type of the method is not part of a signature. As an example, consider the declarations in the following code snippet:

```
public int getAgeInMonths() {
    ...
}

public float getAgeInMonths() {
    ...
}
```

The signatures for both of these methods are identical. The return type is not used. If we attempt to declare both methods in the Employee class we will get the following syntax error message:

`getAgeInMonths() is already defined in packt.Employee`

Calling methods

The syntax used for invoking methods appears similar to using instance variables. Instance methods will always execute against an object. The normal syntax uses the name of the object followed by a period and then the name of the method and any parameters that are needed. In the following example, the getAgeInMonths method is invoked against the employee reference variable:

```
Employee employee = new Employee();
System.out.println(employee.getAgeInMonths());
```

Static methods are invoked using either the class name or an object. Consider the following declarations for a static variable called entityCode:

```
public class Employee {
    // static variables
    private static int entityCode;

    public static void setEntityCode(int entityCode) {
        Employee.entityCode = entityCode;
    }
    ...
}
```

Both the method calls in the following code snippet will invoke the same method:

```
Employee employee = new Employee();
employee.setEntityCode(42);
Employee.setEntityCode(42);
```

However, it is not good practice to use a reference variable to invoke a static method. Instead, always use the classname. Attempts to use an object will result in the following syntax warning:

`Accessing static method setEntityCode`

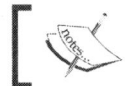

> Static methods are detailed in the *Instance and static class members* section.

A parameter list may be empty if no parameters are passed to a method. In the following simplified method, the age of an employee is returned in months. No parameters are passed to the method and an integer is returned. The method is simplified as the actual value would need to consider the current date and the date of birth for the employee:

```
public int getAgeInMonths() {
    int months = age*12;
    return months;
}
```

Overloading methods

Multiple methods with the same name are permitted in Java. This provides a convenient technique for implementing methods that differ in argument types. Overloaded methods all have the same method name. The methods are differentiated in that each overloaded method must have a unique signature. Signatures are discussed in the earlier *Signature* section. Recall that the return type of the method is not part of a signature.

The following code snippet illustrates the overloading of a method:

```
int max(int, int);
int max(int, int, int);   // Different number of parameters
int max(int ...);         // Varying number of arguments
int max(int, float);      // Different type of parameters
int max(float, int)       // Different order of parameters
```

Classes, Constructors, and Methods

Care must be taken when calling an overloaded method, as the compiler may be unable to determine which method to use. Consider the following declarations of the max method:

```
class OverloadingDemo {

    public int max(int n1, int n2, int n3) {
        return 0;
    }

    public float max(long n1, long n2, long n3) {
        return 0.0f;
    }

    public float max(float n1, float n2) {
        return 0.0f;
    }
}
```

The following code sequence illustrates situations that will give the compiler problems:

```
int num;
float result;
OverloadingDemo demo = new OverloadingDemo();
num = demo.max(45, 98, 2);
num = demo.max(45L, 98L, 2L);        // assignment issue
result = demo.max(45L, 98L, 2L);
num = demo.max(45, 98, 2L);          // assignment issue
result = demo.max(45, 98, 2L);
result = demo.max(45.0f, 0.056f);
result = demo.max(45.0, 0.056f);     // Overload problem
```

The second and fourth assignment statements will match the method call with the three long argument method. This is expected for the second one. For the fourth assignment, only one of the arguments is long but it uses the three long argument method anyway. The problem with these assignments is that the method returns long and not int. It is unable to assign a float value to a int variable without a loss of precision, as indicated by the following syntax error message:

```
possible loss of precision
  required: int
  found:    float
```

[210]

The last assignment cannot find an acceptable overloaded method. The following syntax error message results:

```
no suitable method found for max(double,float)
```

> Closely related to overloading is the process of overriding a method. With overriding, the signatures of two methods are identical but they reside in different classes. This topic is covered in the *Overriding Methods* section in *Chapter 7, Inheritance and Polymorphism*.

Accessors/mutators

An accessor method is one that reads or accesses a variable of a class. A mutator method is one that modifies a variable of a class. These methods are usually public and the variables are normally declared as private. This is an important part of data encapsulation. Private data is hidden from the user but access is provided through methods.

There is a consistent naming convention that you should use with accessor and mutator methods. This convention uses the private member variable name as a base and prefixes the base with either a get or set prefix. The get method returns the value of the variable while the set method takes an argument that is assigned to the private variable. In both methods, the member variable name is capitalized.

This approach is illustrated for the private age field of the Employee class:

```java
public class Employee {
    ...
    private int age;
    ...
    public int getAge() {
        return age;
    }

    public void setAge(int age) {
        this.age = age;
    }
}
```

Notice that the return type of getAge is int and is also the parameter type of the setAge method. This is the standard format of accessors and mutators. Accessor methods are commonly referred to as getters and mutator methods are referred to as setters.

Private data is frequently encapsulated by making it private and providing public methods to access it. Fields that have private or non-existent setters are referred to as **read-only fields**. Fields that have private or non-existent getters are referred to as **write-only fields**, but are not as common. The chief reason for getters and setters is to restrict access and provide additional processing of the fields.

For example, we might have a `getWidth` method that returns the width of a `Rectangle` class. However, the value returned may be dependent on the unit of measure being used. It may return a value depending on whether another unit of measurement field is set to inches, centimeters or pixels. In a security conscious environment, we might want to restrict what can be read or written dependent upon the user or perhaps the time of day.

Instance and static class members

There are two types of variables or methods:

- Instance
- Static

An instance variable is declared as a part of the class and is associated with an object. A static variable is declared in the same way, except that it is preceded by the `static` keyword. When an object is created, it has its own set of instance variables. However, all objects share a single copy of static variables.

Sometimes, it makes sense to have a single variable that can be shared and accessed by all instances of a class. When used with a variable, it is called a **class variable** and is local to the class itself.

Consider the following `Employee` class:

```java
public class Employee {
    // static variables
    private static int minimumAge;

    // instance variables
    private String name;
    private int zip;
    private int age;

    ...
}
```

Each `Employee` object will have its own copies of the `name`, `zip`, and `age` variables. All `Employee` objects may share the same `minimumAge` variable. The use of a single copy of a variable insures that all of the class can access and use the same value, and space is conserved.

Consider the following code sequence:

```
Employee employee1 = new Employee();
Employee employee2 = new Employee();
```

The following diagram illustrates the allocation of the two objects in the heap. Each object has its own set of instance variables. The single static variable is shown allocated above the heap in its own special area of memory:

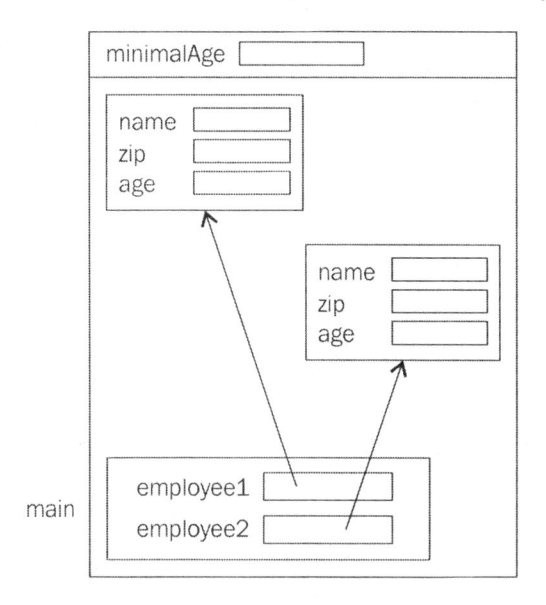

There is only one copy of each method for a class regardless of whether the method is an instance method or a static method. A static method is declared the same way as an instance method, except that the `static` keyword precedes the declaration of the method. The static `setMinimumAge` method, in the following code snippet, illustrates the declaration of a static method:

```
public static void setMinimumAge(int minimumAge) {
    Employee.minimumAge = minimumAge;
}
```

All instance methods must execute against an object. It is not possible to execute against the name of a class as it is possible with a static method. Instance methods are designed to access or modify instance variables. As such, it needs to execute against an object that possesses instance variables. If we attempt to execute an instance method against a classname, shown as follows:

```
Employee.getAge();
```

It will result in the following syntax error message:

`non-static method getAge() cannot be referenced from a static context`

A static method can execute against either an object or the class name. Static methods may not access instance variables or call instance methods. As a static method can execute against a classname this means that it can execute even though there may not be any objects in existence. If there are no objects, then there cannot be any instance variables. Thus, static methods cannot access instance variables.

A static method may not call an instance method. If it were able to access an instance method, then it would indirectly be able to access an instance variable. As there may not be any objects in existence, the calling of instance methods by a static method is not allowed.

An instance method may access a static variable or call a static method. Static variables are always present. Thus, there is no reason why an instance method should not be able to access static variables and methods.

The following table summarizes the relationship between static/instance variables and methods:

	Variable		Method	
	Instance	Static	Instance	Static
Instance method	✓	✓	✓	✓
Static method	✗	✓	✗	✓

Summary

In this chapter we examined many of the important aspects of a class. This included how memory is managed when an instance of a class is created, the initialization process, and how methods can be invoked to use a class.

There are several issues relevant to both constructors and methods. These were discussed before the details of constructors and methods were detailed and included the use of the `this` keyword, passing parameters, and signatures. Constructors and various initialization techniques were illustrated including the order that these initializations take place. The declaration of methods was also discussed including how to overload methods.

We also examined the difference between instance and static, variables, and methods. Throughout the chapter we illustrated how memory is allocated.

Now that we have learned about the basics of classes we are ready to address the topics of inheritance and polymorphism, as discussed in the next chapter. In that chapter we will expand upon how memory is allocated, the initialization sequence, and introduce new topics, such as overriding methods.

Certification objectives covered

The certification objectives covered in this chapter include:

- Creating methods with arguments and return values
- Applying the `static` keyword to methods and fields
- Creating an overloaded method
- Differentiating between default and user-defined constructors
- Applying access modifiers
- Applying encapsulation principles to a class
- Determining the effect upon object references and primitive values when they are passed into methods that change the values

Test your knowledge

1. Which of the following declares a method that takes a float and an integer returns an array of integers?

    a. `public int[] someMethod(int i, float f) { return new int[5]; }`

    b. `public int[] someMethod(int i, float f) { return new int[]; }`

    c. `public int[] someMethod(int i, float f) { return new int[i]; }`

    d. `public int []someMethod(int i, float f) { return new int[5]; }`

2. What happens if you try to compile and run the following code:

    ```
    public class SomeClass {
       public static void main(String arguments[]) {
          someMethod(arguments);
       }
       public void someMethod(String[] parameters) {
          System.out.println(parameters);
       }
    }
    ```

    a. Syntax error – `main` is not declared correctly.

    b. Syntax error – the variable parameters cannot be used as it is in the `println` method.

    c. Syntax error – `someMethod` needs to be declared as static.

    d. The program will execute without errors.

3. Which of the following statements about overloaded methods are true?

    a. Static methods cannot be overloaded.

    b. The return value is not considered when overloading a method.

    c. Private methods cannot be overloaded.

    d. An overloaded method cannot throw exceptions.

4. Given the following code, which of the following statements are true?

    ```
    public class SomeClass {
       public SomeClass(int i, float f) { }
       public SomeClass(float f, int i) { }
       public SomeClass(float f) { }
       public void SomeClass() { }
    }
    ```

[216]

a. A syntax error will occur because void cannot be used with a constructor.

b. A syntax error will occur because the first two constructors are not unique.

c. The class does not have a default constructor.

d. No syntax errors will be generated.

5. Which of the following keywords cannot be used when declaring a class?

   a. public
   b. private
   c. protected
   d. package

6. Assuming that the following classes are in the same package, which statements are true?

```
class SomeClass {
   void method1() { }
   public void method2( { }
   private void method3( { }
   protected void method4() { }
}

class demo [
   public void someMethod(String[] parameters) {
      SomeClass sc = new SomeClass();
      sc.method1();
      sc.method2();
      sc.method3();
      sc.method41();
   }
}
```

a. `sc.method1()` will generate a syntax error.

b. `sc.method2()` will generate a syntax error.

c. `sc.method3()` will generate a syntax error.

d. `sc.method4()` will generate a syntax error.

e. No syntax errors will be generated.

7. What is the output of the following code?

```
public static void main(String args[]) {
    String s = "string 1";
    int i = 5;
    someMethod1(i);
    System.out.println(i);
    someMethod2(s);
    System.out.println(s);
}

public static void someMethod1(int i) {
    System.out.println(++i);
}
public static void someMethod2(String s) {
    s = "string 2";
    System.out.println(s);
}
```

    a. 5 5 string 2 string 1
    b. 6 6 string 2 string 2
    c. 5 5 string 2 string 2
    d. 6 5 string 2 string 1

7
Inheritance and Polymorphism

In this chapter, we will examine how Java supports several important object-oriented concepts including inheritance and polymorphism. When the term "inheritance" comes to mind, we think of that rich uncle who will leave us his vast fortune. Or we say that she has her mother's eyes. In programming terms we talk about classes and how they are related to each other. The terms, "parent class" and "child class", are used to describe an inheritance relationship between classes where the class has access to the capabilities of the parent class.

> There are several terms used to designate a parent class and a child class. You may see the parent class referred to as the super class or base class. The child class may be called the subclass or the derived class. In this chapter we will use the terms **base class** and **derived class**.

The base class typically has methods that implement common functionality needed by that class and the classes that are derived from that class. For example, we may have a *person* class that represents an individual. It may have methods that allow us to maintain the name or age of a person. We may create other classes that represent different types of people — butcher, baker, or candle-stick maker. These different types of people have different capabilities above and beyond those capabilities we define for the *person* class.

When we implement a baker for example, that class might have a method called *cook* that is used to cook something. However, the baker still has a name and an age. Instead of re-implementing the code to support the modification of the name or age we would prefer to re-use the code we developed for the person class. The process of doing this is called inheritance.

Inheritance and Polymorphism

Inheritance allows us to re-use the capabilities of the base class. This, in turn, promotes the re-use of software and can make the developer more productive.

We will also explain how polymorphism is supported in Java. This concept is important and assists in making an application more maintainable. Polymorphism is the result of overriding methods of a base class. Overriding is similar to overloading but it uses the same signature as a base class method.

Polymorphism is frequently used in conjunction with abstract classes. An abstract class is one which cannot be instantiated. That is, it is not possible to create an instance of that class. While we cannot create an instance of an abstract class, we can create an instance of a class derived from the abstract class. This capability can enhance the structure of an application.

With inheritance comes the need to invoke constructors of a base class. We will examine the approach used by Java to control the initialization sequence. Also, the idea of determining the type of a class and casting between classes of an inheritance hierarchy becomes important in some situations.

The last topic addressed in this chapter is concerned with the organization of memory as it relates to inheritance. Understanding how memory is organized and handled will deepen your understanding of the language and assist in debugging applications.

Inheritance

Inheritance is concerned with the relationship between two classes—the base class and the derived class. In this section we will cover the following:

- Implementing a subclass
- Using the `protected` keyword
- Overriding methods
- Using the `@Override` annotation
- Using the `final` keyword with classes
- Creating abstract methods and classes

The use of constructors and inheritance is covered in the *The super keyword and constructors* section.

When inheritance occurs, the derived class inherits all of the methods and attributes of the base class. However, it can only access the public and protected members of the class. It cannot access the private members of the class.

When a method is added to a derived class that has the same signature and accessibility of a base class method, the method is said to override the base class method. This allows the derived class to redefine the meaning of that method. The examples in this chapter will use a `Employee` base class and a `SalaryEmployee` class that is derived from the base class.

Implementing a subclass

A class is implemented through the use of the `extends` keyword, followed by the base classname. In the following example, the `Employee` base class is defined:

```
class Employee {
   // Implementation of Employee class
}
```

The `SalaryEmployee` class can be derived from the base class `Employee`, as shown in the following code snippet:

```
class SalaryEmployee extends Employee  {
   // Implementation of SalaryEmployee class
}
```

Inheritance is used extensively throughout Java libraries. For example, applets are created by extending the `Applet` class.

> A significant part of becoming a skilled Java programmer is learning to find, understand, and use those classes found in libraries relevant to your application's domain.

In the following example, the `HelloWorldApplet` class extends and inherits all of the methods and properties of this class. In this case, the `paint` method is overridden by `HelloWorldApplet`:

```
import java.awt.Graphics;

public class HelloWorldApplet extends java.applet.Applet {

   public void paint (Graphics g) {
      g.drawString ("Hello World!", 5, 15);
   }

}
```

It is possible, and entirely desirable, for a base class to have one or more derived classes. In the case of the `Employee` base class, we might create not only a `SalaryEmployee` class but also a `HourlyEmployee` class. They will share the common capabilities of the base class and yet contain their own unique capabilities.

Let's examine the `Employee` base class and the `SalaryEmployee` class more carefully. First, let's start with a more detailed implementation of the `Employee` class, as shown in the following code snippet:

```java
class Employee {
   private String name;
   private int zip;
   private int age;
   ...
   public int getAge() {
      return age;
   }
   public void setAge(int age) {
      this.age = age;
   }
   ...
}
```

In this implementation we have only included a private `age` instance variable and a getter and setter method for it. In the following `SalaryEmployee` class, we have not added any fields:

```java
class SalaryEmployee extends Employee {
   // Implementation of SalaryEmployee class
}
```

However, even though we haven't added anything new to the `SalaryEmployee` class, it has the capabilities of the base class. In the following sequence we create an instance of both classes and use their methods:

```java
public static void main(String[] args) {
   Employee employee1 = new Employee();
   SalaryEmployee employee2 = new SalaryEmployee();

   employee1.setAge(25);
   employee2.setAge(35);

   System.out.println("Employee1 age: " +
      employee1.getAge());
   System.out.println("Employee2 age: " +
      employee2.getAge());
}
```

When the code is executed, we get the following output:

`Employee1 age: 25`

`Employee2 age: 35`

As the `getAge` and `setAge` methods were public we can use them with the `SalaryEmployee` class even though we haven't defined new versions. However, if we attempt to access the private `age` variable, as shown in the following code snippet, we will get a syntax error:

```
employee2.age = 35;
```

The syntax error generated is as follows:

`age has private access in Employee`

In the *A review of scope* section, we will explore scoping and inheritance in more depth.

> Java does not support multiple inheritance between classes. That is, a derived class cannot extend more than one class. A derived class can extend one and only one class. However, Java does support multiple inheritance between interfaces.

Using the protected keyword

In the previous example, we determined that we cannot access private members from an instance variable, `employee2.age` in the example. We cannot access it from methods or constructors of the derived class either. In the following implementation of the `SalaryEmployee` class, we attempt to initialize the `age` variable in its default constructor:

```java
public class SalaryEmployee extends Employee{

    public SalaryEmployee() {
        age = 35;
    }

}
```

The syntax error is as follows:

`age has private access in Employee`

Inheritance and Polymorphism

However, any base class members declared as public can be accessed from either a member method, or constructor of the derived class, or by an instance variable referencing the derived class.

There are situations where a member variable should be accessible from a derived class constructor, or method, but not from its instance variables. We may want to restrict access to the member at a finer level than either public or private. In the case of the `age` variable, we may trust the derived class to use the variable correctly but we may not trust the user of the instance variable. Using a protected field limits where the field can be modified within the application and where potential problems can be introduced.

This is where the protected access modifier comes in. By using the keyword, `protected`, with a base class member, we restrict access to that member. It is accessible from only within the base class or from derived class constructors or methods.

In the following implementation of the `Employee` class, the `age` variable is declared as protected:

```
class Employee {
    protected int age;
    ...

    public int getAge() {
        return age;
    }

    public void setAge(int age) {
        this.age = age;
    }

    ...
}
```

The `age` variable is now accessible from the `SalaryEmployee` class, as illustrated below where it is initialized:

```
public SalaryEmployee() {
    age = 35;
}
```

This initialization does not generate a syntax error. However, we still cannot access the `age` variable from an instance reference variable. The following code will still result in a syntax error assuming that the class the statement resides in is not in the same package as the `SalaryEmployee` class. This is explained in the *A review of scope* section:

```
employee2.age = 35;
```

The `protected` keyword can also be used with methods. Its use with methods furthers your ability to control access to class members. As an example, the following implementation of the `Employee` class uses the `protected` keyword with the `setAge` method:

```
class Employee {
   protected int age;
   ...
   public int getAge() {
      return age;
   }
   protected void setAge(int age) {
      this.age = age;
   }
   ...
}
```

This means that any user of the class can use the `getAge` method but only the base class, classes in the same package, or derived classes can access the `setAge` method.

Overriding methods

While methods of a base are automatically available for use in a derived class, the actual implementation may not be correct for a derived class. Consider the use of a method to compute the pay of an employee. A `computePay` method in the `Employee` class may simply return a base amount, as shown in the following code snippet:

```
class Employee {
   private float pay = 500.0f;
   public float computePay() {
      return pay;
   }
   ...
}
```

This example is based on the float data type which is not necessarily the best data type for representing currency values. The `java.math.BigDecimal` class is better suited for this purpose. We used the float data type here to simplify the example.

However, for a derived class such as `HourlyEmployee`, the `computePay` method is not correct. This situation can be rectified by overriding the `computePay` method, as shown in the following simplified `HourlyEmployee` implementation:

```java
public class HourlyEmployee extends Employee{
    private float hoursWorked;
    private float payRate;

    public HourlyEmployee() {
       this.hoursWorked = 40.0f;
       this.payRate = 22.25f;
    }

    public float computePay() {
       return hoursWorked * payRate;
    }

}
```

An overridden method possesses two basic characteristics:

- Has an identical signature as a base class method
- Is found in the derived class

The signature of a class is composed of its name, the number of parameters, the type of the parameters, and the order of the parameters. This is discussed in more detail in the *Signature* section of *Chapter 6, Classes, Constructors, and Methods*.

The terms overloading and overriding are easily confused. The following table summarizes the key differences between these terms:

Characteristic	Overloading	Overriding
Method name	Same	Same
Signature	Different	Same
Class	Same class	In a derived class

Let's look at the use of the `computePay` method. In the following sequence, the `computePay` method is executed against the `employee1` and `employee3` instance variables:

```
Employee employee1 = new Employee();
HourlyEmployee employee3 = new HourlyEmployee();

System.out.println("Employee1 pay: " +
   employee1.computePay());
System.out.println("Employee3 pay: " +
   employee3.computePay());
```

The output will be as follows:

Employee1 pay: 500.0

Employee3 pay: 890.0

The `computePay` method of the `Employee` base class is executed against the `employee1` reference variable while the `computePay` method of `HourlyEmployee` is executed against the `employee3` reference variable. The **Java Virtual Machine (JVM)** determines which method to use as the program executes. This is actually an example of polymorphic behavior, which we will address in the *Polymorphism* section.

In a more complex hierarchy of classes, intermediate classes may not override a method. For example, if a `SupervisorEmployee` class was derived from the `SalaryEmployee` class it is not necessary for the `SalaryEmployee` class to implement the `computePay` method. The `SupervisorEmployee` class can override the `computePay` method of `Employee`, whether or not its immediate base class overrode it.

The @Override annotation

One Java language design issue concerns method overriding. The problem is that a developer may intend to override a method but may not, due to a simple error in the method declaration. In the following attempt to override the `computePay` method, however, the method name is misspelled:

```
public float computPay() {
     return hoursWorked * payRate;
}
```

Inheritance and Polymorphism

While it may (or may not be) obvious that the method is misspelled, the developer may not notice the mistake. Using the previous example:

```
Employee employee1 = new Employee();
HourlyEmployee employee3 = new HourlyEmployee();

System.out.println("Employee1 pay: " +
   employee1.computePay());
System.out.println("Employee3 pay: " +
   employee3.computePay());
```

The program will still execute but will not generate the expected output as shown below:

```
Employee1 pay: 500.0
Employee3 pay: 500.0
```

Notice that the base class' `computePay` method is used in both cases. This is because the `computePay` method was invoked instead of the misspelled `computPay` method. As the `HourlyEmployee` class no longer has a `computePay` method, the JVM uses the base class method. Obviously, this is not what was intended.

It can be hard to immediately spot these types of errors. To assist in preventing these types of mistakes, we can use the `@Override` annotation with the method as follows:

```
@Override
public float computPay() {
    return hoursWorked * payRate;
}
```

This annotation informs the compiler to make sure that the method that follows, actually overrides a base class method. In this case, it does not because the name of the method is misspelled. When this happens, a syntax error is generated indicating that there is a problem. The syntax error message is as follows:

```
method does not override or implement a method from a supertype
```

When the spelling of the method is corrected, the syntax error message will go away.

As the name annotation implies, an annotation is a way of adding additional information to parts of a Java application that can be processed at a later time. In the case of the `@Override` annotation at compile time, a check is made to verify that overriding actually took place. Annotations can be used for other purposes, such as marking a method as deprecated.

> It is a good practice to always use the `@Override` annotation with overridden methods.

Using the final keyword with classes

The `public`, `abstract`, and `final` keywords can be used when declaring a class. The `public` keyword specifies the scope of the class, as will be explained in the *A review of scope* section. The use of the `abstract` keyword is covered in the next section, *Abstract methods and classes*. When the `final` keyword is used before the `class` keyword, it signifies that the class cannot be extended. It will be the last class in that branch of the inheritance hierarchy.

In the following example, the `Employee` class is designated as a final class. While it will not make sense to make the `Employee` class final for this chapter's examples, it does illustrate the syntax required to make a class final:

```
final class Employee {
    ...
}
```

By restricting other classes from extending a class, you can be assured that the intended operation of the class will not be compromised by a derived class' overriding a base class method. If it is well implemented, this can result in a more reliable foundation from which to build applications.

The `java.lang.String` class is an example of a class found in the core JDK that is defined as final. It is not possible to extend this class or modify its behavior. This means that developers worldwide can use the class and not worry about the possibility of inadvertently using a derived class instead of the `String` class.

The `final` keyword can also be used with a method definition. When used in this context, it means that the method cannot be overridden in a derived class. This provides more flexibility than making a class final. The developer can specify those methods that may be overridden and those that cannot be overridden.

The following illustrates making the `getAge` method final in the `Employee` class:

```
public class Employee {
    ...
    public final int getAge() {
        return age;
    }
}
```

If we attempt to override the method in a derived class, such as the `SalaryEmployee` class, we will get the following error message:

```
getAge() in SalaryEmployee cannot override getAge() in Employee
  overridden method is final
```

Abstract methods and classes

Abstract classes are useful in the design of an object-oriented inheritance hierarchy. They are typically used to force a derived class to implement a specific set of methods. The base class and/or one or more methods of the class are declared as abstract. An abstract class cannot be instantiated. A non-abstract class, in contrast, must have all of the abstract methods in its hierarchy tree, if any, implemented.

The following example illustrates how to make the Employee class abstract. In this example, there are no abstract methods but the abstract keyword was used to designate the class as abstract:

```
public abstract class Employee {
    ...
}
```

As the Employee class has no abstract methods, none of the derived classes will be forced to implement any additional methods. The above definition effectively has no effect on the previous examples in this chapter.

The next definition of the Employee class makes the computePay method abstract. Notice that the method does not have a body but is terminated with a semicolon instead:

```
public abstract class Employee {
    ...
    public abstract float computePay();
    ...
}
```

All classes that are immediately derived from the Employee class must implement the abstract method or they, themselves, will become abstract. If they elect not to implement the computePay method, the class must be declared as abstract.

When we declare a method as abstract, we are forced to use the abstract keyword with the class. An abstract class can also possess non-abstract methods.

In complex hierarchies you may find a mixture of non-abstract and abstract classes. For example, in the java.awt package you will find that the non-abstract Container class extends the abstract Component class that extends the non-abstract Object class. An abstract class may be introduced at any level in a hierarchy to meet the needs of the library.

Abstract classes can have final methods but they cannot be declared as final. That is, the `final` keyword cannot be used as a modifier of an abstract class or method. If this was possible, it would be impossible to extend the class. As it is abstract it could never be instantiated and, thus, would be useless. However, an abstract class can have final methods. These methods must be implemented in that abstract class. The class can still be extended, but the final methods cannot be overridden.

Polymorphism

Polymorphism is a key object-oriented programming concept but it can be hard to understand initially. The primary purpose of using polymorphism is to make applications more maintainable. When we speak of polymorphism, we typically say that a method exhibits polymorphic behavior.

> A method is said to be polymorphic if the behavior of the method is dependent on the object it is executing against.

Suppose we want to draw something. Each class may have a method called `draw` that it can use to draw itself. For example, a circle class might have a draw method that actually draws itself as a circle. A person class might have a draw method that displays an image of that person. The signature of the methods is identical.

Thus, if we apply the `draw` method against different objects of different classes with the same ultimate base class, then the result of the draw will differ depending on whether we are applying the `draw` method against a circle or a person. That is polymorphic behavior.

By designing our application to use polymorphism, we can add new classes that have a draw method and integrate them into our application a lot easier than previously possible in non-object oriented programming languages.

When an instance of an object is created, the object goes through a series of initialization steps as detailed in the *Java initialization sequence* section in *Chapter 6, Classes, Constructors, and Methods*. This applies to objects that are derived from base classes also. Java memory management is dynamic and automatic. When the `new` keyword is used, it automatically allocates memory from the heap.

In Java, a reference to a base class and any of its derived classes can be assigned to a base class reference variable. This is possible because of the way memory is allocated for base and derived classes. In a derived class, the instance variables of the base class are allocated first, followed by the instance variables of the derived class. When a base class reference variable is assigned to a derived class object, it sees the base class instance variables that it expects plus the "extra" derived class instance variables.

Inheritance and Polymorphism

Let's use the following definitions of the `Employee` and `SalaryEmployee` classes:

```
public class Employee {
    private String name;
    private int age;

    ...

}


public class SalaryEmployee extends Employee {
    private float stock;
    ...
}
```

In the following example, assigning either a `Employee` or a `SalaryEmployee` object to the base class reference makes sense from the perspective of the reference variable because it expects to see the instance variables `name` and `age`. We can assign a new `Employee` object to the `employee` variable, as shown in the following code snippet:

```
Employee employee;
employee = new Employee();
```

This is also illustrated in the following diagram:

```
employee ──────▶ ┌──────┬───┐
                 │ name │   │
                 │ age  │   │
                 └──────┴───┘
```

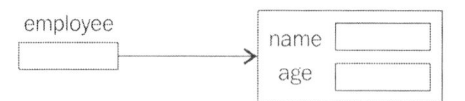

We can also assign a new `SalaryEmployee` object to the `employee` variable using the following code. Notice that in the previous figure, and in this figure, the `employee` reference variable points to a `name` and a `age` field in that order. The `employee` reference variable expects a `Employee` object consisting of a `name` field and then an `age` field and that is what it sees:

```
employee = new SalaryEmployee();
```

This scenario is depicted in the following diagram:

```
employee ──────▶ ┌───────┬───┐
                 │ name  │   │
                 │ age   │   │
                 │ stock │   │
                 └───────┴───┘
```

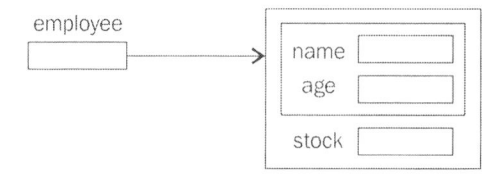

If the following code snippet is executed, based upon the previous declarations of the `Employee` and `SalaryEmployee` classes, the `computePay` method of `SalaryEmployee` will be executed, and not that of the `Employee` class:

```
Employee employee = new SalaryEmployee();
System.out.println(employee.computePay());
```

The `computePay` method is said to be polymorphic in relation to the object it is running against. If the `computePay` method ran against an `Employee` object, the `Employee`'s `computePay` method would execute.

A reference to a derived object can be assigned to that class' object reference variable or to any of that class' base classes. The advantage of polymorphic behavior can be better understood through the next example. Here, the sum of all of the employee's pay from the `employees` array is calculated:

```
Employee employees[] = new Employee[10];
float sum = 0;

// initialize array
employees[0] = new Employee();
employees[1] = new SalaryEmployee();
employees[2] = new HourlyEmployee();
...

for(Employee employee : employees) {
    sum += employee.computePay();
}
```

The `computePay` method is executed against each element of the array. Based on the object it is executing against, the appropriate `computePay` method is invoked. If a new class is derived from the `Employee` class, such as a `SalesEmployee` class, the only modification needed to make the summation process work is to add a `SalesEmployee` object to the array. No other changes are needed. The result is a more maintainable and extensible application.

The allocation of the memory for derived classes helps explain how polymorphism works. We can assign a reference to a `SalaryEmployee` to either a `SalaryEmployee` reference variable or to an `Employee` reference variable. This is illustrated in the following code sequence:

```
Employee employee1 = new Employee();
SalaryEmployee employee2 = new SalaryEmployee();
employee1 = new SalaryEmployee();
employee1 = employee2;
```

Inheritance and Polymorphism

All of the above assignments are legal. It is possible to assign a derived class object to a base class reference variable because the base class reference variable is actually pointing to the memory whose first part contains the base class instance variables. This is illustrated in the following diagram, where each stack reflects the cumulative effect of the four assignment statements:

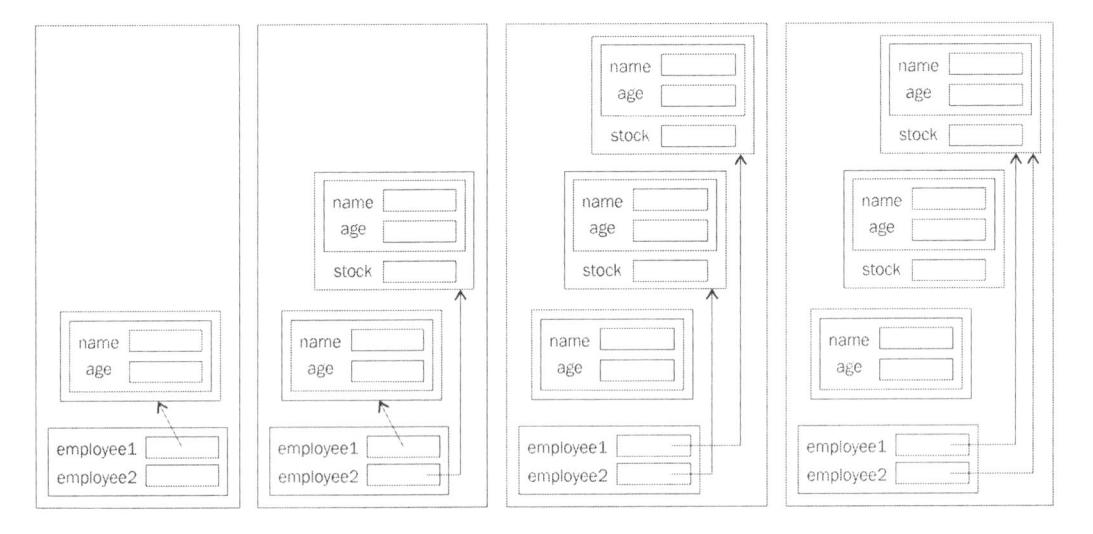

Notice that some objects are no longer referenced by the application. These objects are eligible for garbage collection. At some point, if needed, they will be returned to the heap.

Managing classes and objects

This section is concerned with a number of issues relating to the general management of classes and objects. It includes the:

- Creation and initialization of objects
- Accessing methods of a base class
- Determining the type of objects
- Use of the `Object` class
- Casting objects
- Controlling the scope of classes and members

The super keyword and constructors

As we saw in the *Using the this Keyword* section of *Chapter 6, Classes, Constructors, and Methods*, the `this` keyword refers to the current object. It can be used to:

- Access instance variables
- Pass the current object to a method
- Return the current object from a method

The `super` keyword is used in a complementary fashion within a derived class. It is used to:

- Call the base class constructor
- Access an overridden method in the base class

Calling a base class constructor

Let's examine its use in calling base class constructors. When a derived class object is created, it is initialized by calling its constructor. The use of constructors is covered in the *Constructors* section of *Chapter 6, Classes, Constructors, and Methods*. However, before a derived class constructor is executed, the base class constructor is invoked. This results in the base class being initialized before the derived class. This is particularly important should the derived class use any of the base class methods in the initialization sequence.

The invocation of the base class' default constructor occurs automatically unless we use the `super` keyword to invoke an alternative base class constructor. The following is an implementation of the `Employee` class which defines two constructors—a default constructor and a three argument constructor:

```java
public class Employee {
   private String name;
   private int zip;
   private int age;
   ...
   public Employee() {
      this("Default name", 12345, 21);
   }
   public Employee(String name, int age, int zip) {
      this.name = name;
      this.zip = zip;
      this.age = age;
   }
   ...
}
```

Inheritance and Polymorphism

Notice the use of the `this` keyword to call the three argument constructor. The following is a partial implementation of the `SalaryEmployee` class. Only a default constructor is defined:

```
public class SalaryEmployee extends Employee {
   private int age;
   ...
   public SalaryEmployee() {
      age = 35;
   }
}
```

In this example, the default constructor of the `Employee` class is automatically invoked. However, we can make this invocation explicit by using the `super` keyword followed by a set of open and close parentheses, as shown below:

```
public SalaryEmployee() {
   super();
   age = 35;
}
```

In both cases, member variables of the `Employee` base class in the `SalaryEmployee` object are initialized as specified in the base class constructor.

> If we explicitly use the `super` keyword to invoke a base class constructor, it must be the first line of the derived class' constructor. The `this` keyword must be the first statement in a constructor to invoke another constructor of the same class. Both of these keywords cannot be used in the same constructor to invoke another constructor.

However, there are situations where we may wish to invoke some constructor other than the default base class constructor. To do this we use the `super` keyword as the first statement in a derived class constructor and supply a list of parameters that match those of another base class constructor. In the following example, the four argument constructor of `SalaryEmployee` calls the base class' three argument constructor:

```
public SalaryEmployee(String name, int age, int zip,
      float stock) {
   super(name, age, zip);
   this.stock = stock;
}
```

If we could not choose the base class constructor, then we would need to explicitly call the appropriate setter methods to initialize the base class variables after the base class' default constructor has executed. This is illustrated in the following code snippet:

```java
public SalaryEmployee(String name, int age, int zip,
        float stock) {
    this.setName(name);
    this.setAge(age);
    this.setZip(zip);
    this.stock = stock;
}
```

This approach is not the preferred approach. It is best to allow the base class to initialize its own member variables. A derived class is not always in a position to determine how they should be initialized, and if they are completely hidden the derived class may not even be aware of their existence.

> If a constructor invokes base class methods, these methods should be declared as final. Otherwise, a derived class that overrides them could adversely affect the initialization sequence.

Accessing an overridden method in the base class

We can also use the `super` keyword to access the overridden method of a base class method. For example, it is always a good idea to override the `toString` method of a class to return a string that represents that instance of the class. One possible implementation of this method for the `Employee` class is provided in the following code snippet:

```java
public class Employee {
    ...
    @Override
    public String toString() {
        return "Name: " + this.name +
            "  Age: " + this.age;
    }
}
```

An implementation for the `SalaryEmployee` class is shown in the following code snippet, which uses the base class getter methods to return the name and age:

```java
public class SalaryEmployee extends Employee {
    ...
    @Override
    public String toString() {
        return "Name: " + this.getName() +
               "  Age: " + this.getAge() +
               "  Stock: " + this.stock;
    }
}
```

However, this implementation is awkward, as it requires invocation of the getter methods. Another problem with this approach is that each derived class may provide a different representation of the base class variables resulting in possible confusion on the part of the user of this method.

Ideally, we would simply call the base class' `toString` method in this example to get the string representation of the base class. However, calling the `toString` method from within the derived class's `toString` method results in a recursive call. That is, the runtime system thinks we are calling the current method. This is illustrated in the following code snippet:

```java
public class SalaryEmployee extends Employee {
    ...
    @Override
    public String toString() {
        // Results in a recursive call to the current method
        return toString() + "  Stock: " + this.stock;
    }
}
```

We can avoid these issues by using the `super` keyword to invoke a base class method. This is done by prefixing the name of the base class method with the `super` keyword and a period, as shown in the following code snippet:

```java
public class SalaryEmployee extends Employee {
    ...
    @Override
    public String toString() {
        return super.toString() + "  Stock: " + this.stock;
    }
}
```

The effect of using the `super` keyword is illustrated in the next code sequence:

```
Employee employee1 = new Employee("Paula", 23, 12345);
SalaryEmployee employee2 =
   new SalaryEmployee("Phillip", 31, 54321, 32);

System.out.println(employee1);
System.out.println(employee2);
```

The output will appear as follows:

```
Name: Paula   Age: 23
Name: Phillip   Age: 31   Stock: 32.0
```

Notice that the `toString` method is not explicitly invoked in the `println` method. When an object reference is used within a `print` or a `println` method, the `toString` method is automatically invoked if no other methods are used with the reference variable.

Unlike having to use the `super` keyword as the first statement in a constructor to invoke a base class constructor, when used to invoke a derived class method the `super` keyword can be used anywhere. It does not have to be used in the same overridden method.

In the example that follows, the `display` method invokes the base class's `toString` method:

```
public class SalaryEmployee extends Employee {
   ...
   public void display() {
      System.out.println("Employee Base Data");
      System.out.println(super.toString());
      System.out.println("SalaryEmployee Data");
      System.out.println("Stock: " + this.stock);
   }
}
```

Here, the `display` method is invoked against the `employee2` reference variable:

```
SalaryEmployee employee2 = new SalaryEmployee();
employee2.display();
```

The resulting output follows:

```
Employee Base Data
Name: Phillip   Age: 31
SalaryEmployee Data
Stock: 32.0
```

Inheritance and Polymorphism

It is not possible to call a base method above the current base class. That is, assuming an inheritance hierarchy of Employee - SalaryEmployee - Supervisor, a base class method of the Employee class cannot be called directly from a Supervisor method. The following code will result in a syntax error message:

```
super.super.toString();   //illegal
```

Determining the type of an object

There are times when it is useful to know an object's class. There are a couple of ways to determine its type. The first approach is to get the classname using the Class class. The second approach is to use the instanceof operator.

There is actually a class in Java named Class and it is found in the java.lang package. It is used to obtain information regarding the current object. For our purposes we will use its getName method to return the name of the class. First we obtain an instance of Class using the getClass method. This method is a member of the Object class. The following illustrates this approach:

```
Employee employee1 = new Employee();
SalaryEmployee employee2 = new SalaryEmployee();

Class object = employee1.getClass();
System.out.println("Employee1 type: " + object.getName());
object = employee2.getClass();
System.out.println("Employee2 type: " + object.getName());
```

When this sequence is executed, we get the following output. In this example, the class names are prefixed with the name of their package. All classes developed for this book were placed within the **packt** package:

```
Employee1 type: packt.Employee
Employee2 type: packt.SalaryEmployee
```

While knowing the name of the class may be useful in some situations, the instanceof operator is frequently more useful. We can use the operator to determine whether one object is an instance of a class. This is demonstrated in the following example where we determine the type of classes referenced by the employee1 and employee2 variables:

```
System.out.println("Employee1 is an Employee: " +
    (employee1 instanceof Employee));
System.out.println("Employee1 is a SalaryEmployee: " +
    (employee1 instanceof SalaryEmployee));
System.out.println("Employee1 is an HourlyEmployee: " +
    (employee1 instanceofHourlyEmployee));
```

[240]

```
System.out.println("Employee2 is an Employee: " +
    (employee2 instanceof Employee));
System.out.println("Employee2 is a SalaryEmployee: " +
    (employee2 instanceof SalaryEmployee));
```

This sequence displays a true or false value based on the operator's operands. The output is as follows:

```
Employee1 is an Employee: true
Employee1 is a SalaryEmployee: false
Employee1 is an HourlyEmployee: false
Employee2 is an Employee: true
Employee2 is a SalaryEmployee: true
```

The Object class

The `Object` class is found in the `java.lang` package. This class is the ultimate base class of all Java classes. If a class does not explicitly extend a class, Java will automatically extend that class from the `Object` class. To illustrate this, consider the following definition of the `Employee` class:

```
public class Employee {
    // Implementation of Employee class
}
```

While we did not explicitly extend the `Object` class, it is extended from the `Object` class. To verify this, consider the following code sequence:

```
Employee employee1 = new Employee();
System.out.println("Employee1 is an Object: " +
    (employee1 instanceof Object));
```

The output is as follows:

```
Employee1 is an Object: true
```

The application of the `instanceof` operator confirms that the `Employee` class is ultimately an object of `Object`. The above definition of the `Employee` class has the same effect as if we had derived it from the `Object` explicitly, as shown in the following code snippet:

```
public class Employee extends Object {
    // Implementation of Employee class
}
```

Using a common base class in Java guarantees that all classes have common methods. The `Object` class possesses several methods that most classes may need, as listed in the following table:

Method	Meaning
clone	Produces a copy of the object.
equals	Returns true if two objects are "equal".
toString	Returns a string representation of the object.
finalize	Performed before an object is returned to the heap manager.
getClass	Returns a `Class` object that provides additional information about an object.
hashCode	Returns a unique hash code for the object.
notify	Used in thread management.
notifyAll	Also used in thread management.
wait	Overloaded method used in thread management.

> When a new class is created it is always a good idea to override the `toString`, `equals`, and `hashCode` methods.

> Before an object can be cloned, its class must implement the `java.lang.Cloneable` interface. The `clone` method is protected.

Casting objects

In Java, we are able to cast an object to a different class, other than that of the original object. The cast can be up or down the hierarchy chain. When we cast a derived class object to a base class reference variable, it is called **upcasting**. When we cast a base class object to a derived class reference variable, it is called **downcasting**. Let's start with the following declaration where `Employee` is the base class of `SalaryEmployee`:

```
Employee employee1;
SalaryEmployee employee2;
```

The following example illustrates upcasting. An instance of the derived class, `SalaryEmployee` is assigned to the base class reference variable `employee1`. This is legal and is an important part of polymorphic behavior:

```
employee1 = new SalaryEmployee();
```

The next statement attempts to perform downcasting. An instance of the base class is being assigned to the derived class reference variable. This statement will result in a syntax error:

```
employee2 = new Employee(); // Syntax error
```

However, the syntax error can be avoided by using the cast operator as follows:

```
employee2 = (SalaryEmployee) new Employee();
```

But, a `ClassCastException` exception, shown as follows, will be thrown when the preceding statement is executed:

java.lang.ClassCastException: packt.Employee cannot be cast to packt.SalaryEmployee

Upcasting is possible because the derived object contains everything that the base class has, plus something more. Downcasting is not a good idea as the reference variable expects an object with more capabilities than is supplied.

Notice, with upcasting, that the methods available to the reference variable are those of the base class and not the derived class. Even though the reference variable points to the derived class object, it can only use the base class methods because that's what we've told the Java compiler the object is. This is illustrated in the following statement where we try to use the derived class' `setStock` method:

```
employee1.setStock(35.0f);
```

The following syntax error will be generated for this statement:

cannot find symbol
symbol: method setStock(float)
** location: variable employee1 of type Employee**

A review of scope

Scope refers to when a variable is visible and can be accessed. In earlier chapters we learned how the `public` and `private` keywords are used to control the scope of member variables. In the *Using the protected keyword* section of this chapter, we explored how the `protected` keyword works. However, the declaration of a member variable does not require the use of any of these keywords. When modifiers are not used, the variable declaration is called **package-private**. As the name implies, the scope of the variable is restricted to those classes in the same package.

Inheritance and Polymorphism

We also need to consider the use of the `public` keyword when used with a class definition. If a class is declared as public, it is visible to all classes. If no declaration is used, its visibility is limited to the current package. The class is said to have package-private visibility.

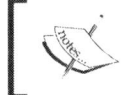

> The use of the `private` and `protected` keywords cannot be used with a class definition unless the class is an inner class. An inner class is a class that is declared within another class.

The following table summarizes the scope of access modifiers as applied to class member variables and methods:

Modifier	Class	Package	Derived Class	Other
public	✓	✓	✓	✓
private	✓	✗	✗	✗
protected	✓	✓	✓	✗
none	✓	✓	✗	✗

Let's also consider the following package/class arrangement, which provides a more detailed look into the scoping rules:

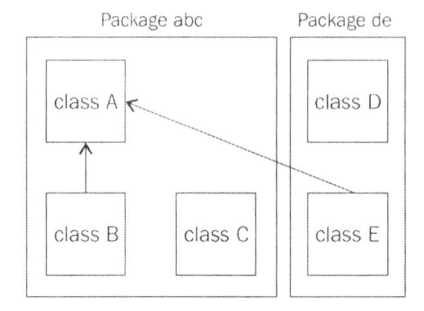

Assume that class A has the following declarations:

```
public class A {
    public int v1;
    private int v2;
    protected int v3;
    int v4;
}
```

The following table summarizes the scoping rules for these declarations. These rules apply to both variables and methods declared in class A. It is slightly different from the previous table as it illustrates the placement of a derived class in a different package. Thus, the access permissions in the protected row appear to be different from the previous table:

Variable	A	B	C	D	E
`public int v1;`	✓	✓	✓	✓	✓
`private int v2;`	✓	✗	✗	✗	✗
`protected int v3;`	✓	✓	✓	✗	✓
`int v4;`	✓	✓	✓	✗	✗

It may be necessary to declare an instance of class A in some of these classes in order to have access to the instances variables of A. For example, in class D the following code is needed to access class A:

```
A a = new A();
a.v1 = 35;
...
```

> In general, use the most restrictive access that makes sense. This will improve the reliability of the application by avoiding situations where accidental access to a member results in unanticipated consequences.

Summary

In this chapter we have studied inheritance and polymorphic behavior, as defined by Java. We have examined how objects are allocated in memory to gain a more comprehensive understanding of how polymorphism and constructors work. The use of the `this` and `super` keywords were examined with regards to their use in constructors and derived class methods. In addition, abstract classes were examined, along with how they impact on polymorphic behavior.

The `protected` and `final` keywords were covered. We saw how the `final` keyword can affect inheritance and overriding methods. The `protected` keyword allowed us to better control the access to information in a derived class.

Inheritance and Polymorphism

The management of classes and objects was addressed including the organization of classes in a package and how to obtain information about objects using the `Class` class. The use of package-protected members was introduced. Also covered was the use of casting with classes.

In the next chapter we will cover the important topic of exception handling. Knowing how to properly use exception handling will enable you to create more robust and maintainable programs.

Certification objectives covered

The certification objectives addressed in this chapter include:

- Implementing inheritance
- Developing the code that demonstrates the use of polymorphism
- Differentiating between the type of a reference and the type of an object
- Determining when casting is necessary
- Using `super` and `this` to access objects and constructors
- Using abstract classes and interfaces

Test your knowledge

1. Which set of statements result in `ClassB` and `ClassC` being derived from `ClassA`?

    a. `class ClassB extends ClassA {}`
    b. `class ClassB extends ClassC {}`
    c. `class ClassA extends ClassB {}`
    d. `class ClassC extends ClassB {}`
    e. No combination will work

2. Which of the following must be true for a method to support polymorphism?

    a. The method must override a base class method
    b. The method must overload a base class method
    c. The method's class must extend a base class that has the overridden method
    d. The method must execute against a base class reference variable

3. What method is used to determine the type of an object?

    a. `isType`

    b. `typeOf`

    c. `instanceof`

    d. `instanceOf`

4. Which of the following are valid casts?

    a. `num1 = num2;`

    b. `num1 = (int)num2;`

    c. `num1 = (float)num2;`

    d. `num1(int) = num2;`

5. Given the following class definitions:

```
public class ClassA {
    public ClassA() {
        System.out.println("ClassA constructor");
    }

    public void someMethod() {
        System.out.println("ClassA someMethod");
    }
}

class ClassB extends ClassA {
    public ClassB() {
        System.out.println("ClassB constructor");
    }

    public void someMethod() {
        // comment
        System.out.println("ClassB someMethod");
    }
    public static void main(String args[]) {
        ClassB b = new ClassB();
        b.someMethod();

    }
}
```

What statement is needed at the comment line to generate the following output:

```
ClassA constructor
ClassB constructor
ClassA someMethod
ClassB someMethod
```

    a. `super();`

    b. `super().someMethod;`

    c. `super.someMethod();`

    d. `someMethod();`

    e. None of the above

6. Which of the following statements are true?

    a. An abstract class must use the abstract keyword when declared

    b. An abstract class must have one or more abstract methods

    c. An abstract class cannot extend a non-abstract class

    d. An abstract class cannot implement an interface

8
Handling Exceptions in an Application

Exceptions are objects that are thrown by an application or the **Java Virtual Machine (JVM)** when an error of some sort occurs. Java provides a wide range of predefined exceptions and allows the developer to declare and create their own exception classes.

While there are a number of ways of classifying exceptions, one scheme classifies them into three types:

- Program errors
- Improper use of code
- Resource-related failures

Program errors are internal flaws in a code sequence. The programmer may or may not be able to do much about these types of errors. For example, a common exception is `NullPointerException`. This is frequently a result of not properly initializing or assigning a value to a reference variable. This type of error can be hard to avoid and anticipate when first writing a piece of code. However, once detected, the code can be revised to correct the situation.

Code may be improperly used. Most libraries are designed to be used in a specific manner. They may expect data to be organized in one way and if the user of the library fails to follow the format, an exception may be thrown. For example, the parameter of a method may not be structured as expected by the method or may be of the wrong type.

Some errors are related to resource failure. When the underlying system is not able to satisfy the program's needs, a resource type of exception can occur. For example, a failure in the network may prevent the program from executing properly. This type of error may require re-executing the program at a later time.

A traditional approach to handling exceptions is to return an error code from a procedure. For example, a function may normally return a zero if it executed without an error. If an error did not occur, a non-zero value would be returned. The problem with this approach is that the calling of the function may either:

- Be unaware that the function returns an error code (for example, C's `printf` function)
- Forget to check for an error
- Ignore the error completely

When the error is not caught, the continued execution of the program can lead to unpredictably and possibly disastrous consequences.

An alternative to this method is to "catch" errors. Most modern block structured languages such as Java use this approach. This technique requires less coding and is more readable and robust. When a routine detects an error, it "throws" an exception object. The exception object is then returned to the caller which then catches and handles the error.

Exceptions should be caught for a number of reasons. Failure to deal with exceptions can result in the application failing, or ending up in an invalid state with incorrect output. It is always a good idea to maintain a consistent environment. Also, if you open a resource, such as a file, you should always close the resource when you are done except for the most trivial programs.

The exception handling mechanisms available in Java allow you to do this. When a resource is opened, it can be closed even if an exception occurs in the program. To accomplish this task, a resource is opened in a `try` block and closed in a `catch` or `finally` block. The `try`, `catch`, and `finally` blocks constitute the core of the exception handling mechanism used in Java.

Exception types

Java has provided an extensive set of classes to support exception handling in Java. An exception is an instance of a class derived directly, or indirectly, from the `Throwable` class. Two predefined Java classes are derived from `Throwable`—`Error` and `Exception`. From the `Exception` class is derived a `RuntimeException` class. As we will see shortly, programmer-defined exceptions are normally derived from the `Exception` class:

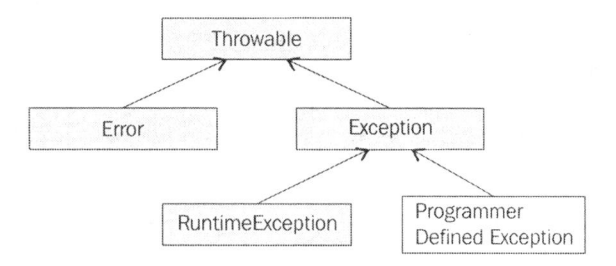

```
                    Throwable
                   /         \
               Error        Exception
                            /         \
                  RuntimeException   Programmer
                                     Defined Exception
```

There are numerous pre-defined errors that are derived from the `Error` and `RuntimeException` classes. There is little that a programmer will do with the exceptions derived from the `Error` object. These exceptions represent problems with the JVM and normally can't be recovered. The `Exception` class is different. The two classes that derive from the `Exception` class support two types of exceptions:

- **Checked**: These are exceptions that need to be dealt with in the code
- **Unchecked**: These are exceptions that do not need to be dealt with in the code

Checked exceptions include all exceptions derived from the `Exception` class and are not derived from the `RuntimeException` class. These must be handled in code or the code will not compile cleanly, resulting in compile-time errors.

Unchecked exceptions are all other exceptions. They include exceptions, such as division by zero and array subscripting errors. These do not have to be caught but like the `Error` exceptions, if they are not caught, the program will terminate.

We can create our own exception classes. When we do, we need to decide whether to create a checked or unchecked exception. A general rule of thumb is to declare the exception as an unchecked exception if the client code cannot do anything to recover from the exception. Otherwise, if they can handle it, make it a checked exception.

> The users of a class do not have to account for unchecked exceptions that can result in the program terminating, if the client program does not ever deal with them. A checked exception requires the client to either catch the exception or explicitly pass it up the call hierarchy.

Exception handling techniques in Java

There are three general techniques we can use when dealing with exceptions in Java:

- Traditional `try` block
- The new "try-with-resources" block introduced in Java 7
- Pass the buck

The third technique is used when the current method is not the appropriate place to handle the exception. It allows the exception to be propagated higher into the sequence of method calls. In the following example, `anotherMethod` may encounter some condition where it may throw `IOException`. Instead of dealing with the exception in `someMethod`, the `throws` keyword in the `someMethod` definition results in the exception being passed to the code that called this method:

```java
public void someMethod() throws IOException {
    ...
    object.anotherMethod(); // may throw an IOException
    ...
}
```

The method will skip all of the remaining lines of code in the method and immediately return to the caller. Uncaught exceptions are propagated to the next higher context until they are caught or they are thrown from `main`, where an error message and stack trace will be printed.

Stack trace

The `printStackTrace` is a method of the `Throwable` class that will display the program stack at that point in the program. It is used automatically when an exception is not caught or can be called explicitly. The output of the method pinpoints the line and method that caused the program to fail. You have seen this method in action before, whenever you had an unhandled runtime exception. The method is automatically called when an exception is not handled.

The `ExceptionDemo` program illustrates the explicit use of the method:

```java
public class ExceptionDemo {

    public void foo3() {
        try {
            ...
            throw new Exception();
        }
```

```
        catch (Exception e) {
            e.printStackTrace();
        }
    }
    public void foo2() { foo3(); }
    public void foo1() { foo2(); }

    public static void main(String args[]) {
        new ExceptionDemo().foo1();
    }
}
```

The output is shown as follows:

```
java.lang.Exception
    at ExceptionDemo.foo3(ExceptionDemo.java:8)
    at ExceptionDemo.foo2(ExceptionDemo.java:16)
    at ExceptionDemo.foo1(ExceptionDemo.java:20)
    at ExceptionDemo.main(ExceptionDemo.java:25)
```

Using Throwable methods

The `Throwable` class possesses a number of other methods that can provide more insight in to the nature of the exception. To illustrate the use of many of these methods we will use the following code sequence. In this sequence we attempt to open a non-existent file and examine the exception thrown:

```
private static void losingStackTrace(){
    try {
        File file = new File("c:\\NonExistentFile.txt");
        FileReader fileReader = new FileReader(file);
    }
    catch (FileNotFoundException e) {
        e.printStackTrace();

        System.out.println();
        System.out.println("---e.getCause(): " +
                e.getCause());
        System.out.println("---e.getMessage(): " +
                e.getMessage());
        System.out.println("---e.getLocalizedMessage(): " +
                e.getLocalizedMessage());
        System.out.println("---e.toString(): " +
                e.toString());
    }
}
```

Handling Exceptions in an Application

Due to the nature of some IDEs, an application's standard output and standard error output can be interleaved. For example, the execution of the above sequence may result in the following output. You may not or may see the interleaving in your output. The dashes in front of the output are used to help see the interleaving behavior:

```
java.io.FileNotFoundException: c:\NonExistentFile.txt (The system cannot find the file specified)
---e.getCause(): null
---e.getMessage(): c:\NonExistentFile.txt (The system cannot find the file specified)
    at java.io.FileInputStream.open(Native Method)
---e.getLocalizedMessage(): c:\NonExistentFile.txt (The system cannot find the file specified)
---e.toString(): java.io.FileNotFoundException: c:\NonExistentFile.txt (The system cannot find the file specified)
    at java.io.FileInputStream.<init>(FileInputStream.java:138)
    at java.io.FileReader.<init>(FileReader.java:72)
    at packt.Chapter8Examples.losingStackTrace(Chapter8Examples.java:64)
    at packt.Chapter8Examples.main(Chapter8Examples.java:57)
```

The methods used in this example are summarized in the following table:

Method	Meaning
getCause	Returns the cause of the exception. If it cannot be determined it returns null.
getMessage	Returns a detailed message.
getLocalizedMessage	Returns a localized version of the message.
toString	Returns the string version of the message.

Notice that the first line of the `printStackTrace` method is the output of the `toString` method.

The `getStackTrace` method returns an array of `StackTraceElement` objects where each element represents a line of the stack trace. We can duplicate the effect of the `printStackTrace` method with the following code sequence:

```
try {
    File file = new File("c:\\NonExistentFile.txt");
    FileReader fileReader = new FileReader(file);
}
catch (FileNotFoundException e) {
    e.printStackTrace();
```

```
        System.out.println();
        StackTraceElement traces[] = e.getStackTrace();
        for (StackTraceElement ste : traces) {
            System.out.println(ste);
        }
    }
```

When executed we get the following output:

```
java.io.FileNotFoundException: c:\NonExistentFile.txt (The system cannot find the file specified)
    at java.io.FileInputStream.open(Native Method)
    at java.io.FileInputStream.<init>(FileInputStream.java:138)
    at java.io.FileReader.<init>(FileReader.java:72)
    at packt.Chapter8Examples.losingStackTrace(Chapter8Examples.java:64)
    at packt.Chapter8Examples.main(Chapter8Examples.java:57)

java.io.FileInputStream.open(Native Method)
java.io.FileInputStream.<init>(FileInputStream.java:138)
java.io.FileReader.<init>(FileReader.java:72)
packt.Chapter8Examples.losingStackTrace(Chapter8Examples.java:64)
packt.Chapter8Examples.main(Chapter8Examples.java:57)
```

The traditional try-catch block

The traditional technique to handle exceptions uses a combination of a try, catch, and finally blocks. A try block is used to surround code that might throw exceptions and is followed by zero or more catch blocks and then, optionally, by a single finally block.

The catch blocks are added after a try block to "catch" exceptions. The statements in the catch block provide blocks of code to "handle" the error. A finally clause can optionally be used after the catch blocks. It is guaranteed to execute even if code within a try or a catch block throws or does not throw an exception.

> However, a finally block will not execute if the System.exit method is invoked in a try or catch block.

The following sequence illustrates the use of these blocks. Within the try block, a line is read in and an integer is extracted. Two catch blocks are used to handle the exceptions that might be thrown:

```
try {
   inString = is.readLine();
   value = Integer.parseInt (inString);
   ...
}
catch (IOException e) {
   System.out.println("I/O Exception occurred");
}
catch (NumberFormatException e) {
   System.out.println("Bad format, try again...");
}
finally {
   // Perform any necessary clean-up action
}
```

One of two types of errors is possible in this code sequence:

- Either an error will occur trying to read a line of input or
- An error will occur trying to convert the string to an integer

The first catch block will catch IO errors and the second catch block will catch conversion errors. Only one catch block is ever executed when an exception is thrown.

An error may, or may not, occur. Regardless, the finally block will execute either after the try block completes or after a catch block executes. The `finally` clause is guaranteed to run and generally contains "clean-up" code.

Using the try-with-resource block

The use of the previous technique can be cumbersome when multiple resources are opened and a failure occurs. It can result in multiple try-catch blocks that become hard to follow. In Java 7, the try-with-resources block was introduced to address this situation.

The advantage of the try-with-resources block is that all resources opened with the block are automatically closed upon exit from the block. Any resources used with the try-with-resources block must implement the interface `java.lang.AutoCloseable`.

We will illustrate this approach by creating a simple method to copy one file to another. In the following example, one file is opened for reading and the other is opened for writing. Notice how they are created between the `try` keyword and the block's open curly brace:

```
try (BufferedReader reader = Files.newBufferedReader(
    Paths.get(new URI("file:///C:/data.txt")),
      Charset.defaultCharset());
    BufferedWriter writer = Files.newBufferedWriter(
      Paths.get(new URI("file:///C:/data.bak")),
      Charset.defaultCharset())) {

  String input;
  while ((input = reader.readLine()) != null) {
    writer.write(input);
    writer.newLine();
  }
} catch (URISyntaxException | IOException ex) {
  ex.printStackTrace();
}
```

Resources to be managed are declared and initialized inside a set of parentheses and are placed between the `try` keyword and the opening curly brace of the try block. The first resource is a `BufferedReader` object that uses the `data.txt` file and the second resource is a `BufferedWriter` object used with the `data.bak` file. The `Paths` class is new to Java 7 and provides improved IO support.

Resources declared with a try-with-resources block must be separated by semicolons otherwise a compile-time error will be generated. More in-depth coverage of the try-with-resources block can be found in The Java 7 Cookbook.

The use of the vertical bar in the catch block is new to Java 7 and allows us to catch multiple exceptions in a single catch block. This is explained in the *Using the | operator in a catch block* section.

Catch statement

The catch statement has exactly one argument. The catch statement will trap the exception if its parameter:

- Exactly matches the exception type
- Is a base of the exception type
- Is an interface that the exception type implements

Only the first catch statement that matches the exception will execute. If no matches are made, the method will terminate and the exception will bubble up to the calling method where it may be handled.

A part of the earlier `try` block is duplicated as follows. The format of the `catch` statement consists of the `catch` keyword followed by a set of open and close parentheses enclosing an exception declaration. The set of parentheses is then followed by zero or more statements in a block statement:

```
try {
   ...
}
catch (IOException e) {
   System.out.println("I/O Exception occurred");
}
catch (NumberFormatException e) {
   System.out.println("Bad format, try again...");
}
```

The process of handling an error is up to the programmer. It may be as simple as displaying an error message or it can be quite complex. The programmer may use the error object to retry the operation or otherwise deal with it. This may involve propagating it back to the calling method in some situations.

Order of the catch blocks

The order in which catch blocks are listed after a try block can be significant. When an exception is thrown, the exception object is compared to the catch blocks in the order that they are listed. The comparison checks to see if the thrown exception is a type of the exception in the catch block.

For example, if a `FileNotFoundException` is thrown, it will match either a catch block that has an `IOException` or a `FileNotFoundException` exception because `FileNotFoundException` is a sub-type of `IOException`. As the comparison is stopped when the first match is found, if the catch block for `IOException` is listed before the catch block for `FileNotFoundException`, the `FileNotFoundException` block will never be executed.

Consider the following hierarchy of exception classes:

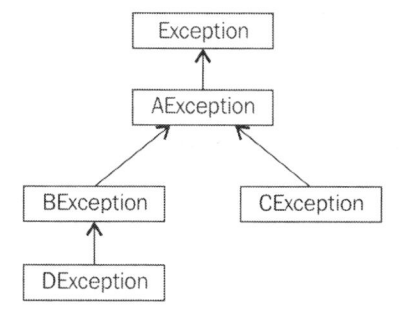

```
        Exception
            ↑
        AException
        ↑       ↑
   BException  CException
        ↑
   DException
```

Given the following code sequence:

```
try {
    ...
}
catch (AException e) {...}
catch (BException e) {...}
catch (CException e) {...}
catch (DException e) {...}
```

If an exception is thrown that is one of these types of exceptions, the AException catch block will always be executed. This is because an AException, BException, CException, or a DException are all of the AException type. The exception will always match the AException exception. The other catch blocks will never be executed.

The general rule is always to list the "most-derived" exceptions first. The following is the correct way of listing the exceptions:

```
try {
    ...
}
catch (DException e) {...}
catch (BException e) {...}
catch (CException e) {...}
catch (AException e) {...}
```

Notice, that it doesn't make any difference with this hierarchy of exceptions whether the BException immediately precedes or follows the CException, as they are at the same level.

Using the | operator in a catch block

Sometimes it is desirable to handle multiple exceptions in the same way. Instead of duplicating the code in each catch block, we can use a vertical bar to permit one catch block to capture more than one exception.

Consider the situation where two exceptions are potentially thrown and are handled in the same way:

```
try {
    ...
}
catch (IOException e) {
    e.printStackTrace();
}
catch (NumberFormatException e) {
    e.printStackTrace();
}
```

A vertical bar can be used to catch two or more exceptions in the same `catch` statement as illustrated in the following code snippet. This can reduce the amount of code needed to handle two exceptions that are handled in the same way.

```
try {
    ...
}
catch (IOException | NumberFormatException e) {
    e.printStackTrace();
}
```

This approach works when more than one exception can be handled in the same way. Keep in mind that the catch block's parameter is implicitly final. It is not possible to assign a different exception to the parameter. The following attempt is illegal and will not compile:

```
catch (IOException | NumberFormatException e) {
    e = new Exception();   // Compile time error
}
```

The finally block

The `finally` block follows a series of `catch` blocks and consists of the `finally` keyword followed by a block of statements. It contains one or more statements that will always be executed to clean up previous actions. The `finally` block will always execute regardless of the existence or non-existence of exceptions. However, if a `try` or `catch` block invokes the `System.exit` method, the program immediately terminates and the `finally` block does not execute.

The purpose of a `finally` block is to close or otherwise handle any resources that were opened within the `try` block. It is always a good practice to close resources after they have been opened and are no longer needed. We will this in the next example.

However, in practice this is often tedious and can be error prone if it is necessary to close multiple resources where the close process may also generate exceptions. In addition, if one resource throws an exception while being opened and another one was not opened, we have to be careful not to attempt to close the second one. As a result, in Java 7 the try-with-resources block has been introduced to address this type of problem. This block was discussed in the *Using the try-with-resource block* section. Here, we will cover the simplified use of the `finally` block.

A simple example of using the `finally` block is shown as follows. In this sequence we will open a file for input and then display its content:

```
BufferedReader reader = null;
try {
   File file1 = new File("c:\\File1.txt");

   reader = new BufferedReader(new FileReader(file1));
   // Copy file
   String line;
   while((line = reader.readLine()) != null) {
      System.out.println(line);
   }
}
catch (IOException e) {
   e.printStackTrace();
}
finally {
   if(reader != null) {
      reader.close();
   }
}
```

The file will be closed regardless of whether an exception was thrown or not. If the file does not exist, a `FileNotFoundException` will be thrown. This will be caught in the `catch` block. Notice how we checked the `reader` variable to make sure it was not null.

In the following example, we open two files and then try to copy one file to another. The finally block is used to close the resources. This illustrates a problem with the finally block when dealing with multiple resources:

```
BufferedReader br = null;
BufferedWriter bw = null;
try {
   File file1 = new File("c:\\File1.txt");
   File file2 = new File("c:\\File2.txt");

   br = new BufferedReader(new FileReader(file1));
   bw = new BufferedWriter(new FileWriter(file2));
   // Copy file
}
catch (FileNotFoundException e) {
   e.printStackTrace();
}
catch (IOException e) {
   e.printStackTrace();
}
finally {
   try {
      br.close();
      bw.close();
   } catch (IOException ex) {
      // Handle close exception
   }
}
```

Notice, that the `close` methods may also throw a `IOException`. We must also handle these exceptions. This may require a more complicated exception handling sequence which can be error prone. In this case, note that the second file will not be closed if an exception is thrown when the first file is closed. In this situation it is better to use the try-with-resources block, as discussed in the *Using the try-with-resources block* section.

> A try block needs either a catch block or a finally block. Without one or both a compile time error will be generated.

Nested try-catch blocks

Exception handling can be nested. This can become necessary when methods are used in a `catch` or `finally` block that also throws exceptions. The following illustrates using a nested `try` block inside of a `catch` block:

```
try {
   // Code that may throw an exception
}
catch (someException e) {
   try {
      // Code to handle the exception
   }
   catch (anException e) {
      // Code to handle the nested exception
   }
}
catch (someOtherException e) {
   // Code to handle the exception
}
```

In the last example of the previous section, we used the `close` method inside a `finally` block. However, the `close` method may throw a `IOException`. As it is a checked exception, we will need to catch it. This results in a `try` block being nested inside of a `finally` block. In addition, when we try to close the `BufferedReader`, a `NullPointerException` will be thrown in the second `try` block because we attempted to execute the close method against the `reader` variable which was never assigned a value.

To complete the previous example, consider the following implementation:

```
finally {
   try {
      br.close();
      bw.close();
   } catch (IOException | NullPointerException e) {
      // Handle close exceptions
   }
}
```

We used the | bar to simplify the capture of both exceptions as detailed in the *Using the | operator in a catch block* section. This is also another example where we may lose the original exception. In this case, the `FileNotFoundException` was lost to a `NullPointerException`. This will be discussed in the *Losing the stack trace* section.

Exception handling guidelines

This section addresses general guidelines for working with exceptions. It is intended to provide examples of how to use exception handling in a more useful and productive manner. While poor techniques may not result in a compile-time error, or an incorrect program, they often reflect a poor design.

Repeating code that threw an exception

When an exception is thrown and then caught we will sometimes want to try and re-execute the offending code. This is not difficult if the code is structured properly.

In this code sequence, errors are assumed to be present when the `try` block is entered. If an error is generated it is caught and handled by the `catch` block. As the `errorsArePresent` is still set to true, the try block will be repeated. However, if no errors occur, at the end of the try block the `errorsArePresent` flag is set to false which will allow the program to execute the while loop and continue executing:

```
boolean errorsArePresent;

...
errorsArePresent = true;
while (errorsArePresent) {
   try {
      ...
      errorsArePresent = false;
   }

   catch (someException e) {
      // Process error
   }

}
```

An assumption is made, in this example, that the code used to process the error will necessitate that the `try` block be re-executed. This may be the case when all we do in the process error code sequence is to display an error message that identifies the error, such as when the user enters a bad filename.

You need to be careful using this approach if the resource needed is not available. This can result in an infinite loop where we check for a resource that is not available, throw an exception, and then do it all over again. A loop counter can be added to specify the number of times we try to handle the exception.

Not being specific in which exception you are catching

When catching an exception, be specific about the one you need to catch. For example, in the following example the generic `Exception` is caught. There is nothing specific that will reveal more useful information about what caused the exception:

```
try {
   someMethod();
} catch (Exception e) {
   System.out.println("Something failed" + e);
}
```

A more useful version follows which catches the actual exception thrown:

```
try {
   someMethod();
} catch (SpecificException e) {
   System.out.println("A specific exception message" + e);
}
```

Losing the stack trace

Sometimes an exception is caught and then a different one is re-thrown. Consider the following method where a `FileNotFoundException` exception is thrown:

```
private static void losingStackTrace(){
   try {
      File file = new File("c:\\NonExistentFile.txt");
      FileReader fileReader = new FileReader(file);
   }
   catch(FileNotFoundException e) {
      e.printStackTrace();
   }
}
```

Assuming that the file does not exist, the following stack trace is generated:

```
java.io.FileNotFoundException: c:\NonExistentFile.txt (The system cannot find the file specified)
   at java.io.FileInputStream.open(Native Method)
   at java.io.FileInputStream.<init>(FileInputStream.java:138)
   at java.io.FileReader.<init>(FileReader.java:72)
   at packt.Chapter8Examples.losingStackTrace(Chapter8Examples.java:49)
   at packt.Chapter8Examples.main(Chapter8Examples.java:42)
```

Handling Exceptions in an Application

We can tell what the precise exception was and where it occurred. Next, consider the use of using the `MyException` class instead of the `FileNotFoundException` exception:

```java
public class MyException extends Exception {
    private String information;

    public MyException(String information) {
        this.information = information;
    }
}
```

If we re-throw the exception, as shown in the following code snippet, we will lose information about the original exception:

```java
private static void losingStackTrace() throws MyException {
    try {
        File file = new File("c:\\NonExistentFile.txt");
        FileReader fileReader = new FileReader(file);
    }
    catch(FileNotFoundException e) {
        throw new MyException(e.getMessage());
    }
}
```

The stack trace that results from this implementation is as follows:

```
Exception in thread "main" packt.MyException
    at packt.Chapter8Examples.losingStackTrace(Chapter8Examples.java:53)
    at packt.Chapter8Examples.main(Chapter8Examples.java:42)
```

Notice, that the details of the actual exception have been lost. In general it is a good idea not to use this approach as information crucial for debugging is lost. Another example of this problem is found in the *Nested try-catch blocks* section.

It is possible to re-throw and preserve the stack trace. To do this we need to do the following:

14. Add a constructor with a `Throwable` object as a parameter.
15. Use this when we want to preserve the stack trace.

The following shows such a constructor added to the `MyException` class:

```java
public MyException(Throwable cause) {
    super(cause);
}
```

In the `catch` block we will use this constructor, as shown below.

```
catch (FileNotFoundException e) {
   (new MyException(e)).printStackTrace();
}
```

We could have thrown the exception. Instead, we use the `printStackTrace` method, shown as follows:

```
packt.MyException: java.io.FileNotFoundException: c:\NonExistentFile.txt
(The system cannot find the file specified)
   at packt.Chapter8Examples.losingStackTrace(Chapter8Examples.java:139)
   at packt.Chapter8Examples.main(Chapter8Examples.java:40)
Caused by: java.io.FileNotFoundException: c:\NonExistentFile.txt (The
system cannot find the file specified)
   at java.io.FileInputStream.open(Native Method)
   at java.io.FileInputStream.<init>(FileInputStream.java:138)
   at java.io.FileReader.<init>(FileReader.java:72)
  at packt.Chapter8Examples.losingStackTrace(Chapter8Examples.java:136)
```

Scoping and block lengths

The scope of any variable declared within a `try`, `catch`, or `finally` block is limited to that block. It is a good idea to limit the scope of a variable as much as possible. In the following example, it is necessary to define the `reader` variable outside of the try and catch blocks because it is needed in the `finally` block:

```
BufferedReader reader = null;
try {
   reader = ...
      ...
}
catch (IOException e) {
   ...
} finally {
   try {
      reader.close();
   }
   catch (Exception e) {
      ...
   }
}
```

Handling Exceptions in an Application

The length of a block should be limited. However, blocks that are too small can result in your code becoming cluttered with the exception handling code. Let's assume there are four methods that can each throw distinct exceptions. If we use separate try blocks for each method we will wind up with code similar to the following:

```
try {
    method1();
}
catch (Exception1 e1) {
    ...
}
try {
    method2();
}
catch (Exception1 e2) {
    ...
}
try {
    method3();
}
catch (Exception1 e3) {
    ...
}
try {
    method4();
}
catch (Exception1 e4) {
    ...
}
```

This is somewhat unwieldy and also presents problems if a `finally` block is needed for each `try` block. A better approach, if these are logically related, uses a single `try` block, shown as follows:

```
try {
    method1();
    method2();
    method3();
    method4();
}
catch (Exception1 e1) {
    ...
}
catch (Exception1 e2) {
    ...
```

```
    }
    catch (Exception1 e3) {
        ...
    }
    catch (Exception1 e4) {
        ...
    }

    finally {
        ...
    }
```

Depending on the nature of the exceptions we can also use a common base class exception or, as introduced in Java 7, we can use the | operator with a single catch block. This is particularly useful if the exceptions can be dealt with in the same way.

However, it is a bad practice to place the entire body of a method in a try/catch block which contains code not related to the exception. It is better to separate the exception handling code from the non-execution handling code if possible.

A general rule of thumb is to keep the length of the exception handling code to a size that can be seen all at once. It is perfectly acceptable to use multiple try blocks. However, make sure that each block contains operations that are logically related. This helps modularize your code and makes it more readable.

Throwing a UnsupportedOperationException object

Methods that are intended to be overridden will sometimes return an "invalid" value to indicate that the method needs to be implemented. For example, in the following code sequence the `getAttribute` method returns `null`:

```
class Base {
    public String getAttribute() {
        return null;
    }
    ...
}
```

However, if the method is not overridden and the base class method is used, problems such as an incorrect result may result, or a `NullPointerException` may be generated, if a method is executed against the return value.

A better approach is to throw an `UnsupportedOperationException` to indicate that the method functionality has not yet been implemented. This is illustrated in the following code sequence:

```
class Base {
    public String getAttribute() {
        throw new UnsupportedOperationException();
    }
    ...
}
```

The method cannot be used successfully until a valid implementation is provided. This approach is used frequently in the Java API. The `java.util.Collection` class' `unmodifiableList` method uses this technique (http://docs.oracle.com/javase/1.5.0/docs/api/java/util/Collections.html#unmodifiableList%28java.util.List%29). Similar effects can be achieved by declaring the method as abstract.

Ignoring exceptions

It is generally a bad practice to ignore exceptions. They are thrown for a reason and if there is something you can do to recover, then you should deal with it. Otherwise, at minimum, you can gracefully terminate your application.

For example, it is common to ignore an `InterruptedException`, as illustrated in the following code snippet:

```
while (true) {
    try {
        Thread.sleep(100000);
    }
    catch (InterruptedException e) {
        // Ignore it
    }
}
```

However, even here something went wrong. For example, if the thread is a part of a thread pool, the pool may be terminating and you should handle this event. Always understand the environment in which your program is running in and expect the unexpected.

Another example of poor error handling is shown in the following code snippet. In this example we ignore the `FileNotFoundException` exception that may be thrown:

```java
private static void losingStackTrace(){
   try {
      File file = new File("c:\\NonExistentFile.txt");
      FileReader fileReader = new FileReader(file);
   }
   catch(FileNotFoundException e) {
      // Do nothing
   }
}
```

This user is not aware that an exception was ever encountered. This is rarely an acceptable approach.

Handle exceptions as late as you can

When an exception is thrown by a method, the user of the method can either deal with it at that point or pass the exception up the call sequence to another method. The trick is to handle the exception at the appropriate level. That level is typically the one that can do something about the exception.

For example, if input is needed from the application's user to successfully handle the exception, then the level best suited for interacting with the user should be used. If the method is part of a library, then it may not be appropriate to assume that the user should be prompted. When we try to open a file and the file does not exist, we don't expect, or want the method we called, to prompt the user for a different file name. Instead, we are more inclined to do it ourselves. In some cases there may not even be a user to prompt as is the case with many server applications.

Catching too much in a single block

When we add catch blocks to an application, we are frequently tempted to use a minimal number of catch blocks by using a base class exception class to capture them. This is illustrated below where the catch block uses the `Exception` class to capture multiple exceptions. Here, we assume that multiple checked exceptions can be thrown and need to be handled:

```java
try {
   ...
}
catch (Exception e) {
   ...
}
```

If they are all handled exactly the same way, this might be alright. However, if they differ in how they should be handled then we need to include additional logic to determine what actually happened. If we ignore the differences, then it can make any debugging process more difficult because we may have lost useful information about the exception. In addition, not only is this approach too coarse but we also catch all `RuntimeException`s which we may not be able to handle.

Instead, it is generally better to catch multiple exceptions in their own catch block, as illustrated in the following code snippet:

```
try {
    ...
}
catch (Exception1 e1) {
    ...
}
catch (Exception1 e2) {
    ...
}
catch (Exception1 e3) {
    ...
}
catch (Exception1 e4) {
    ...
}
```

Logging exceptions

A common practice is to log exceptions even if they are handled successfully. This can be useful in assessing the behavior of an application. Of course, if we cannot handle the exception and need to gracefully terminate the application, error logs can be quite useful in determining what went wrong in the application.

> Log the exception only once. Logging multiple times can confuse whoever is trying to see what happened and create log files larger than they need to be.

Do not use exceptions to control normal logic flow

It is a poor practice to use exceptions where validation should be performed. In addition, throwing an exception uses up additional resources. For example, the `NullPointerException` is a common exception that results when a method is attempted to be executed against a reference variable that has a null value assigned to it. Instead of catching this exception, we should detect this condition and handle it in the normal logic sequence. Consider the following where we catch a `NullPointerException`:

```
String state = ...   // Somehow assigned a null value
try {
   if(state.equals("Ready") { … }
}
catch(NullPointerException e) {
   // Handle null state
}
```

Instead, we should check the value of the state variable before it is used:

```
String state = ...   // Somehow assigned a null value

if(state != null) {
   if(state.equals("Ready") { … }
} else {
   // Handle null state
}
```

The need for the `try` block is eliminated altogether. An alternate approach uses short-circuiting as illustrated in the following code snippet and is covered in the *Short circuit evaluation* section of *Chapter 3, Decision Constructs*. The use of the `equals` method is avoided if the `state` variable is null:

```
String state = ...   // Somehow assigned a null value

if(state != null && state.equals("Ready") {
   // Handle ready state
} else {
   // Handle null state
}
```

Do not try to handle unchecked exceptions

It is usually not worth the effort to deal with unchecked exceptions. Most of these are beyond the control of the programmer and would require significant effort to recover from. For example, a `ArrayIndexOutOfBoundsException`, while the result of a programming error, is not easily dealt with at runtime. Assuming that it would be feasible to modify the array index variable, it may not be clear what new value should be assigned to it or how to re-execute the offending code sequence.

> Never catch `Throwable` or `Error` exceptions. These should never be handled or suppressed.

Summary

Proper exception handling in your program will enhance its robustness and reliability. The `try`, `catch`, and `finally` blocks can be used to implement exception handling within an application. In Java 7, the try-with-resources block has been added which more easily handles the opening and closing of resources. It is also possible to propagate an exception back up the call sequence.

We learned that the order of catch blocks is important in order to properly handle exceptions. In addition, the | operator can be used in a catch block to handle more than one exception in the same way.

Exception handling may be nested to address problems where the code within a catch, or finally block, may also throw an exception. When this happens, the programmer needs to be careful to insure that previous exceptions are not lost and that the new exceptions are handled appropriately.

We also addressed a number of common problems that can occur when handling exceptions. They provided guidance as to avoid poorly structured and error prone code. These included not ignoring exceptions when they occur and to handle exceptions at the appropriate level.

Now that we've learned about the exception handling process, we're ready to wrap up our coverage of the Java certification objectives in the next chapter.

Certification objectives covered

The certification objectives covered in this chapter include:

- Describe what exceptions are used for in Java
- Differentiate among checked exceptions, runtime exceptions, and errors
- Create a try-catch block and determine how exceptions alter normal program flow
- Invoke a method that throws an exception
- Recognize common exception classes and categories

Test your knowledge

1. Which of the following implement checked exceptions?

    a. `Class A extends RuntimeException`
    b. `Class A extends Throwable`
    c. `Class A extends Exception`
    d. `Class A extends IOException`

2. Given the following set of classes:

    ```
    class Exception A extends Exception {}
    class Exception B extends A {}
    class Exception C extends A {}
    class Exception D extends C {}
    ```

    What is the correct sequence of catch blocks for the following `try` block:

    ```
    try {
        // method throws an exception of the above types
    }
    ```

    a. Catch A, B, C, and D
    b. Catch D, C, B, and A
    c. Catch D, B, C, and A
    d. Catch C, D, B, and A

3. Which of the following statements are true?
   a. Checked exceptions are those derived from the Error class.
   b. Checked exceptions should normally be ignored as we cannot not handle them.
   c. Checked exceptions must be re-thrown.
   d. Checked exceptions should be handled at the appropriate method in the call stack.

4. When a method throws a checked exception which of the following are valid responses?
   a. Place the method in a try-catch block.
   b. Do not use these types of methods.
   c. Do nothing as we normally cannot handle checked exceptions.
   d. Use the throws clause on the method which calls this method.

5. What exceptions may the following code generate at runtime?
```
String s;
int i = 5;
try{
    i = i/0;
    s += "next";
}
```
   a. ArithmeticException
   b. DivisionByZeroException
   c. FileNotFoundException
   d. NullPointerException

9
The Java Application

In this chapter we will examine the structure of a Java application from the perspective of packages. The use of the packages and import statements will be covered, along with the underlying directory structure used for packages.

We will also see how Java supports internationalization through the use of locales and resource bundles. An introduction to the use of JDBC will be presented, as well as examining how unused objects are reclaimed. This is commonly referred to as **garbage collection**.

Code organization

The organization of code is an essential part of an application. One can go as far as to say that it is this organization (along with data organization) that determines the quality of an application.

A Java application is organized around packages. Packages contain classes. Classes contain data and code. Code is found in either an initializer list or in a method. This basic organization is shown in the following diagram:

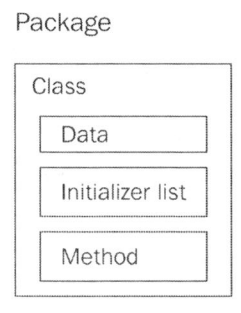

Code can be thought of as being both static and dynamic in nature. The organization of a Java program is structured statically around packages, classes, interfaces, initializer lists, and methods. The only change in this organization comes from different versions of an executing program. However, as a program executes, the myriad of different possible execution paths result in an often complex sequence of execution.

The Java API is organized into many packages of hundreds of classes. New packages and classes are being added on a regular basis making it challenging to keep up with all of the capabilities of Java.

However, as mentioned in the *The Object class* section in *Chapter 7, Inheritance and Polymorphism*, all classes in Java have as a base class—java.lang.Object—either directly or indirectly. In the classes that you defined, if you do not explicitly extend another class, Java will automatically extend this class from the Object class.

Packages

The purpose of a package is to group related classes and other elements together. Ideally, they form a cohesive set of classes and interfaces. A package can consist of:

- Classes
- Interfaces
- Enumerations
- Exceptions

It is natural that classes with similar functionality should somehow be grouped together. Most of Java's IO classes are grouped together in the java.io or java.nio related packages. All of Java's network classes are found in the java.net package. This grouping mechanism provides us with a single logical grouping that is easier to talk about and to work with.

All classes belong to a package. If the package is not specified, then the class belongs to an unnamed default package. This package consists of all the classes in the directory that have not been declared as belonging to a package.

The directory/file organization of packages

To place a class within a package it is necessary to:

- Use the package statement in the class source file
- Move the corresponding .class file to the package directory

The package statement needs to be the first statement in the class' source file. The statement consists of the keyword, `package`, followed by the name of the package. The following example declares that the class `Phone` belongs to the `acme.telephony` package:

```
package acme.telephony;

class Phone {
    ...
}
```

Java source code files are placed in a file with the same name as the class using a `.java` extension. If more than one class is saved in a file, only one class can be declared as public and the file must be named after this public class. The `java.lang` package contains many commonly used classes and is included automatically in every application.

The second requirement is to move the class file to the appropriate package directory. Somewhere on the system there must exist a directory structure that reflects the package name. For example, for a package name `employee.benefits` there needs to be a directory named `employee` that has a subdirectory named `benefits`. All of the class files for the `employee` package are placed in the `employee` directory. All of the class files for the `employee.benefits` package are placed in the `benefits` subdirectory. This is illustrated in the following diagram where the directories and files are located somewhere off in the C drive:

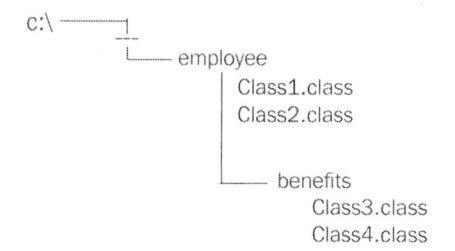

```
c:\
    └── employee
            Class1.class
            Class2.class
            │
            └── benefits
                    Class3.class
                    Class4.class
```

You may also find that a package's directories and classes are compressed into a **Java Archive (JAR)** or `.jar` file. If you look for a specific package structure in a directory system, you may find a JAR file instead. By compressing packages into a JAR file, memory can be minimized. If you find such files, do not unzip them because the Java compiler and JVM expect them to be in a JAR file.

Most IDEs will separate the source files from the class files by placing them in separate directories. This separation makes them easier to work with and deploy.

The import statement

The `import` statement provides information to the compiler in terms of where to find the definition of a class used in the program. There are several considerations regarding the import statement that we will examine:

- Its use is optional
- Using the wildcard character
- Accessing multiple classes with the same name
- The static import statement

Avoiding the import statement

The `import` statement is optional. In the following example, instead of using the `import` statement for the `BigDecimal` class we explicitly use the package name directly in code:

```
private java.math.BigDecimal balance;
   ...
this.balance = new java.math.BigDecimal("0");
```

This is more verbose but it is more expressive. It leaves no doubt that the `BigDecimal` class is the one found in the `java.math` package. However, if we used the class many times in the program then this becomes an annoyance. Normally, the `import` statement is used.

Using the import statement

To avoid having to prefix each class with its package name, the `import` statement can be used to indicate to the compiler where the class can be found. In this example the class, `BufferedReader` of the `java.io` package, can be used without having to prefix the class name with its package name each time it is used:

```
import java.io.BufferReader;
   ...
   BufferedReader br = new BufferedReader();
```

Using the wildcard character

If more than one class needs to be used, and they are found in the same package, an asterisk can be used instead of including multiple import statements, one for each class. For example, if we need to use both the `BufferedReader` and the `BufferedWriter` classes in an application we could use two import statements, as follows:

```
import java.io.BufferedReader;
import java.io.BufferedWriter;
```

By explicitly listing each class, the reader of the code will know immediately where to find the class. Otherwise, the reader might be left guessing from which package a class originates when the wild card character is used with multiple import statements.

While the explicit import of each class is better documentation, the import list can get quite long. Most IDEs support the ability to collapse, or otherwise hide the list.

The alternative approach is to use one import statement with the asterisk, shown as follows:

```
import java.io.*;
```

All of the package's elements can now be used without using the package name. However, this does not mean that the classes of sub-packages can be used in the same way. For example, there are numerous packages that start with `java.awt`. A few of these with some of their elements are shown in the following diagram:

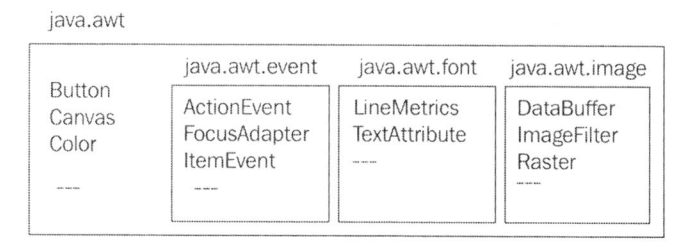

java.awt			
	java.awt.event	java.awt.font	java.awt.image
Button Canvas Color ----	ActionEvent FocusAdapter ItemEvent ----	LineMetrics TextAttribute ----	DataBuffer ImageFilter Raster ----

It may seem the wild card character should include those classes found in these additional packages when used against the "base" package, as shown in the following code:

```
import java.awt.*;
```

However, it imports only those classes in the `java.awt` package and none of the classes in the `java.awt.font` or similar packages. In order to also reference all of the classes of `java.awt.font` also, a second import statement is necessary:

```
import java.awt.*;
import java.awt.font.*;
```

Multiple classes with the same name

As it is possible to have more than one class with the same name in different packages, the import statement is used to specify which class to use. However, the second class will need to explicitly use the package name.

For example, let's assume that we have created a `BigDecimal` class in a `com.company.account` package and we need to use it and the `java.math.BigDecimal` class. We cannot use an import for both classes, as shown in the following code snippet, because this will generate a syntax error to the effect that the names collide.

```
import java.math.BigDecimal;
import com.company.customer.BigDecimal;
```

Instead, we need to either:

- Declare one using the import statement and explicitly prefix the class name of the second when we use it, or
- Do not use the import statement at all and explicitly prefix both classes as we use them

Assuming that we use the `import` statement with the `java.math` class, we use both classes in code, as follows:

```
this.balance = new BigDecimal("0");
com.company.customer.BigDecimal secondary =
    new com.company.customer.BigDecimal();
```

Notice that we had to prefix both usages of `BigDecimal` in the second statement otherwise it would have assumed that the un-prefixed one was in the `java.math` package generating a type mismatch syntax error.

The static import statement

The static import statement can be used to simplify the use of methods. This is commonly used in conjunction with the `println` method. In the following example, we use the `println` method several times:

```
System.out.println("Employee Information");
System.out.println("Name: ");
System.out.println("Department: ");
System.out.println("Pay grade: ");
```

In each case, the classname, `System`, was required. However, if we use the following `import` statement where we added the `static` keyword we will not need to use the `System` classname.:

```
import static java.lang.System.out;
```

The following sequence of code statements achieves the same result:

```
out.println("Employee Information");
out.println("Name: ");
out.println("Department: ");
out.println("Pay grade: ");
```

While this approach saves the time spent in typing, it can be confusing to anyone who does not understand the use of the static import statement.

Garbage collection

Java performs automatic garbage collection. When memory is allocated using the `new` keyword, the memory is obtained from the program heap. This is an area of memory above the program stack. The object allocated is held by the program until the program releases it. This is done by removing all references to the object. Once it is released, the garbage collection routine will eventually run and reclaim the memory allocated by the objects.

The following code sequence illustrates how a `String` object is created. It is then assigned to a second reference variable:

```
String s1 = new String("A string object");
String s2 = s1;
```

The Java Application

At this point, `s1` and `s2` both reference the string object. The following diagram illustrates the memory allocation for `s1` and `s2`:

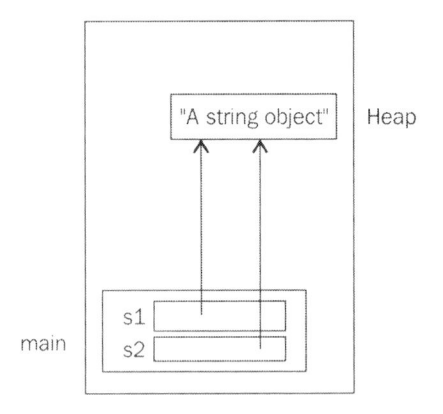

The `new` keyword was used in this situation to ensure that the string object is allocated from the heap. If we had used a string literal, as shown below, the object is allocated to an internal pool as discussed in the *String comparisons* section in *Chapter 2, Java Data Types and Their Usage*:

```
String s1 = "A string object";
```

The next two statements illustrate how the references to the object can be removed:

```
s1 = null;
s2 = null;
```

The following diagram shows the state of the application after these statements have been executed:

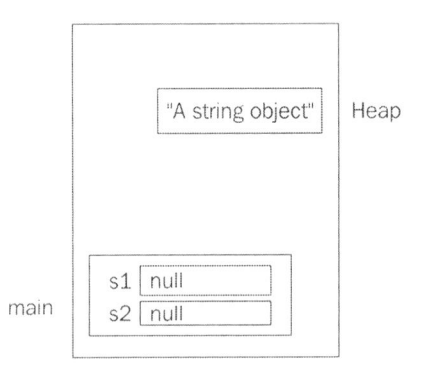

There exists a JVM background thread, which executes periodically to reclaim the unused objects. At some point in the future, the thread will execute. When an object is ready to be reclaimed, the thread will do the following:

- Execute the method's `finalize` method
- Reclaim the memory for re-use by the heap manager

The `finalize` method is not normally implemented by a developer. Its original intent was to correspond to the destructor found in languages such as C++. They were used to perform cleanup activities.

In Java, you should not rely on the method to execute. For small programs, the garbage collection routine may never run as the program may terminate before it has a chance to execute. Over the years, several attempts have been made to provide the ability for a programmer to force the method to execute. None of these attempts have been successful.

Resource bundles and the Locale class

The `Locale` class is used to represent a part of the world. Associated with a locale is a set of conventions concerned with such activities as controlling the way currency or dates are displayed. The use of locales aids in the internationalization of an application. The developer specifies the locale and then uses the locale in various parts of the application.

In addition to the `Locale` class, we can also use resource bundles. They provide a way to customize the appearance based on the locale for data types other than numbers and dates. It is particularly useful when working with strings that change based on the locale.

For example, a GUI application will have different visual components whose text should differ when used in different parts of the world. In Spain, the text and currency should be displayed in Spanish. In China, Chinese characters and conventions should be used. The use of locales can simplify the process of adapting an application to different regions of the world.

In this section we will discuss three approaches used to support internationalization of an application:

- Using the `Locale` class
- Using a property resource file
- Using the `ListResourceBundle` class

Using the Locale class

To illustrate the use of locales we first create an instance of the `Locale` class. This class has a number of predefined locale constants. In the following example we will create a locale for the US and then display the locale:

```
Locale locale;

locale = Locale.US;
System.out.println(locale);
```

The output appears as follows:

en_US

The first part, en_, is short for English. The second part specifies US. If we change the locale to Germany as follows:

```
locale = Locale.GERMANY;
System.out.println(locale);
```

you will get the following output:

de_DE

You can use locales to format currency values. In the following example we have used the static `getCurrencyInstance` method to return an instance of a `NumberFormat` class using the locale for the US. The `format` method is then used to format a double number:

```
NumberFormat currencyFormatter =
    NumberFormat.getCurrencyInstance(Locale.US);
System.out.println(currencyFormatter.format(23.45));
```

The output appears as follows:

$23.45

If we had used the German locale, we would get the following output:

23,45 €

Dates can also be formatted based on a locale. The `DateFormat` class' `getDateInstance` method is used in the following code snippet, with the US locale. The `format` method uses a `Date` object to obtain a string representation of the date, as illustrated in the following code snippet:

```
DateFormat dateFormatter =
    DateFormat.getDateInstance(DateFormat.LONG, Locale.US);
System.out.println(dateFormatter.format(new Date()));
```

The output would be similar to the one that follows:

`May 2, 2012`

In the following code snippet we will use a locale for France:

```
dateFormatter = DateFormat.getDateInstance(
    DateFormat.LONG, Locale.FRANCE);
System.out.println(dateFormatter.format(new Date()));
```

The output of this example is as follows:

`2 mai 2012`

Using resource bundles

Resource bundles are collections of objects organized by the locale. For example, we might have one resource bundle containing strings and GUI components for English speaker and another set for Spanish speakers. These language groups can be further divided into language subgroups such as US versus Canadian English speakers.

A resource bundle can be stored as a file or may be defined as a class. A property resource bundle is stored in a `.properties` file and is restricted to strings. `ListResourceBundle` is a class and can hold strings and other objects.

Using a property resource bundle

A property resource bundle is a file consisting of a set of key-value string pairs where the file name ends with `.properties`. The string key is used to identify a specific string value. For example, a `WINDOW_CAPTION` key can be associated with a string value `Editor`. The following shows the content of a `ResourceExamples.properties` file:

```
WINDOW_CAPTION=Editor
FILE_NOT_FOUND=The file could not be found
FILE_EXISTS=The file already exists
UNKNOWN=Unknown problem with application
```

The Java Application

To access the value in a resource file, we need to create an instance of a `ResourceBundle` class. We can do this by using the `ResourceBundle` class' static `getBundle` method, as shown in the following code snippet. Notice that the resource file name is used as an argument of the method, but it does not include the file extension. If we know the key, we can use it with the `getString` method to return its corresponding value:

```
ResourceBundle bundle = ResourceBundle.getBundle(
    "ResourceExamples");
System.out.println("UNKNOWN" + ":" +
    bundle.getString("UNKNOWN"));
```

The output will appear as follows:

UNKNOWN:Unknown problem with application

We can use the `getKeys` method to obtain an `Enumeration` object. As shown in the following code snippet, the enumeration to display all of the key-value pairs of the file, is used:

```
ResourceBundle bundle = ResourceBundle.getBundle(
    "ResourceExamples");

Enumeration keys = bundle.getKeys();
while (keys.hasMoreElements()) {
   String key = (String) keys.nextElement();
   System.out.println(key + ":" + bundle.getString(key));
}
```

The output of this sequence is as follows:

```
FILE_NOT_FOUND:The US file could not be found
UNKNOWN:Unknown problem with application
FILE_EXISTS:The US file already exists
WINDOW_CAPTION:Editor
```

Notice that the output does not match the order or the contents of the `ResourceExamples.properties` file. The order is controlled by the enumeration. The content is different for the `FILE_NOT_FOUND` and `FILE_EXISTS` keys. This is because it actually used a different file, `ResourceExamples_en_US.properties`. There is a hierarchical relationship between the property resource bundles. The code sequence was executed with a default locale of the US. The system looked for the `ResourceExamples_en_US.properties` file as it represents the strings specific to that locale. Any missing elements in a resource file are inherited from its "base" file.

We will create four different resource bundle files to illustrate the use of resource bundles and the hierarchical relationship between them:

- `ResourceExamples.properties`
- `ResourceExamples_en.properties`
- `ResourceExamples_en_US.properties`
- `ResourceExamples_sp.properties`

These are related to each other hierarchically, as depicted in the following diagram:

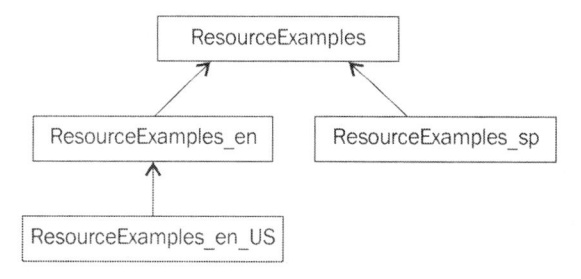

```
                    ResourceExamples
                     ↗          ↖
        ResourceExamples_en   ResourceExamples_sp
               ↑
       ResourceExamples_en_US
```

These files will contain strings for four keys as summarized in the following table:

File	Key	Value
	`WINDOW_CAPTION`	Editor
	`FILE_NOT_FOUND`	The file could not be found
	`FILE_EXISTS`	The file already exists
	`UNKNOWN`	Unknown problem with application
`en`	`WINDOW_CAPTION`	Editor
	`FILE_NOT_FOUND`	The English file could not be found
	`UNKNOWN`	Unknown problem with application
`en_US`	`WINDOW_CAPTION`	Editor
	`FILE_NOT_FOUND`	The US file could not be found
	`FILE_EXISTS`	The US file already exists
	`UNKNOWN`	Unknown problem with application
`sp`	`FILE_NOT_FOUND`	El archivo no se pudo encontrar
	`FILE_EXISTS`	El archivo ya existe
	`UNKNOWN`	Problema desconocido con la aplicación

The en entry is missing a value for the FILE_EXISTS key and the sp entry is missing the WINDOW_CAPTION key. They will inherit the value for the default resource file, as illustrated below for the en locale:

```
bundle = ResourceBundle.getBundle("ResourceExamples",
        new Locale("en"));
System.out.println("en");
keys = bundle.getKeys();
while (keys.hasMoreElements()) {
    String key = (String) keys.nextElement();
    System.out.println(key + ":" + bundle.getString(key));
}
```

The output lists a value for FILE_EXISTS even though it is not found in the ResourceExamples_en.properties file:

en
WINDOW_CAPTION:Editor
FILE_NOT_FOUND:The English file could not be found
UNKNOWN:Unknown problem with application
FILE_EXISTS:The file already exists

The inheritance behavior of these files allows the developer to create a hierarchy of resources files based on a base file name and then extend them by adding a locale suffix. This results in strings being used automatically that are specific for the current locale. If a locale other than the default one is needed, then the specific locale can be specified.

Using the ListResourceBundle class

The ListResourceBundle class is also used to hold resources. Not only can it hold strings, it can also hold other types of objects. However, the keys are still string values. To demonstrate the use of this class, we will create the ListResource class which derives from the ListResourceBundle class as listed below. A static two dimensional array of objects is created containing key-value pairs. Notice the last pair contains an ArrayList. The class' getContents method returns the resources as a two dimensional array of objects:

```
public class ListResource extends ListResourceBundle {

    @Override
    protected Object[][] getContents() {
        return resources;
    }
```

```
          static Object[][] resources = {
             {"FILE_NOT_FOUND", "The file could not be found"},
             {"FILE_EXISTS", "The file already exists"},
             {"UNKNOWN", "Unknown problem with application"},
             {"PREFIXES",new
                  ArrayList(Arrays.asList("Mr.","Ms.","Dr."))}

          };
    }
```

The `ArrayList` created is intended to store various name prefixes. It is created using the `asList` method, which is passed a variable number of string arguments and it returns a `List` to the `ArrayList` constructor.

The following code demonstrates how to use `ListResource`. An instance of `ListResource` is created and then the `getString` method is executed using string keys. For the `PREFIXES` key, the `getObject` method is used:

```
System.out.println("ListResource");
ListResource listResource = new ListResource();

System.out.println(
    listResource.getString("FILE_NOT_FOUND"));
System.out.println(
    listResource.getString("FILE_EXISTS"));
System.out.println(listResource.getString("UNKNOWN"));
ArrayList<String> salutations =
        (ArrayList)listResource.getObject("PREFIXES");
for(String salutation : salutations) {
    System.out.println(salutation);
}
```

The output of this sequence is as follows:

```
ListResource
The file could not be found
The file already exists
Unknown problem with application
Mr.
Ms.
Dr.
```

The Java Application

Using JDBC

JDBC is used to connect to a database and manipulate tables in the database. The process to use JDBC includes the following steps:

1. Connecting to a database
2. Creating an SQL statement to submit to the database
3. Handling the results and any exceptions that may be generated

In Java 7, the use of JDBC has been enhanced with the addition of the try-with-resources block, which simplifies the opening and closing of connections. A detailed explanation of this block is found in the *Using the try-with-resource block* section in Chapter 8.

Connecting to a database

Connecting to a database involves two steps:

1. Loading a suitable driver
2. Establishing a connection

This assumes that a database has been setup and is accessible. In the following examples, we will be using MySQL Version 5.5. MySQL comes with the `Sakila` schema which contains a `customer` table. We will use this table to demonstrate various JDBC techniques.

Loading a suitable driver

First we need to load a driver. JDBC supports a number of drivers, as discussed at http://developers.sun.com/product/jdbc/drivers. Here, we will use the MySQLConnector/J driver. We load the driver using the `Class` class' `forName` method, as shown in the following code snippet:

```
try {
   Class.forName(
            "com.mysql.jdbc.Driver").newInstance();
} catch (InstantiationException |
         IllegalAccessException |
         ClassNotFoundException e) {
   e.printStackTrace();
}
```

The method throws several exceptions which need to be caught.

Note that starting with JDBC 4.0 the above sequence is no longer needed, assuming that the JDBC drivers used support JDBC 4.0. This is the case for the MySQL drivers used with MySQL Version 5.5. This sequence is used here because you will probably run across this approach in older programs.

Establishing a connection

Next, a connection to the database needs to be established. The `java.sql.Connection` represents a connection to a database. The `DriverManager` class' static `getConnection` method will return a connection to a database. Its arguments include:

- A URL representing the database
- A user ID
- A password

The following code sequence will use a try-with-resources block to establish a connection to the database. The first parameter is a MySQL specific connection string. Connection strings are vendor specific:

```
try (Connection connection = DriverManager.getConnection(
        "jdbc:mysql://localhost:3306/", "id", "password")) {
    ...
} catch (SQLException e) {
    e.printStackTrace();
}
```

Creating a SQL statement

Next, we need to create a `Statement` object that will be used to execute a query. The `Connection` class' `createStatement` method will return a `Statement` object. We will add it to the try-with-resources block to create the object:

```
try (Connection connection = DriverManager.getConnection(
        "jdbc:mysql://localhost:3306/", "root", "explore");
     Statement statement = connection.createStatement()) {
    ...
} catch (SQLException e) {
    e.printStackTrace();
}
```

The Java Application

A query string is then formed that will select the first and last name of those customers in the `customer` table whose `address_id` is less than 10. We choose this query to minimize the size of the result set. The `executeQuery` method is used to execute the query and returns a `ResultSet` object that holds the rows of the table matching the selected query:

```java
try (Connection connection = DriverManager.getConnection(
        "jdbc:mysql://localhost:3306/", "root", "explore");
    Statement statement = connection.createStatement()) {
      String query = "select first_name, last_name"
         + " from sakila.customer "
         + "where address_id < 10";
      try (ResultSet resultset =
                      statement.executeQuery(query)) {
         ...
      }
      ...
} catch (SQLException e) {
   e.printStackTrace();
}
```

Handling the results

The last step is to use a while loop to iterate through the result set and display the rows returned. In the following example the `next` method will advance from row to row in the `resultset`. The `getString` method returns the value corresponding to the method's argument that specifies the column to be accessed:

```java
try (Connection connection = DriverManager.getConnection(
        "jdbc:mysql://localhost:3306/", "root", "explore");
    Statement statement = connection.createStatement()) {
      String query = "select first_name, last_name"
         + " from sakila.customer "
         + "where address_id < 10";
      try (ResultSet resultset =
                      statement.executeQuery(query)) {
         while (resultset.next()) {
            String firstName =
                      resultset.getString("first_name");
            String lastName =
                      resultset.getString("last_name");
            System.out.println(firstName + " " + lastName);
         }
      }
} catch (SQLException e) {
   e.printStackTrace();
}
```

The output is as follows:

```
MARY SMITH
PATRICIA JOHNSON
LINDA WILLIAMS
BARBARA JONES
ELIZABETH BROWN
```

JDBC supports the use of other SQL statements such as `update` and `delete`. In addition, it supports the use of parameterized queries and stored procedures.

Summary

In this chapter we have re-examined the overall structure of a Java application. We examined the use of the `import` and `package` statements and discussed the relationship between a package library and its supporting directory/file underpinnings. We learned how to use the asterisk wildcard with the `import` statement. In addition, we saw the use of the static import statement.

We discussed the use of the initializer list and how garbage collection works in Java. This process results in the automatic recovery of objects once they are no longer needed.

The support provided for internationalization was explored starting with the `Locale` class and then with resource bundles. Both, property resource bundles and the `ListResourceBundle` class were covered. We learned how inheritance works with property resource bundles when organized using a consistent naming convention.

Finally, we covered the use of JDBC. We saw how a driver is needed to establish a connection to a database and how to use the `Statement` class to retrieve a `ResultSet` object. This object allowed us to iterate through the rows returned by a `select` query.

Certification objectives covered

The certification objectives covered in this chapter include:

- Defining the structure of a Java class
- Selecting a resource bundle based on locale
- Using the proper JDBC API to submit queries and read results from the database.

Test your knowledge

1. Which of the following will compile without an error?

    a. ```
       package somepackage;
       import java.nio.*;
       class SomeClass {}
       ```

    b. ```
       import java.nio.*;
       package somepackage;
       class SomeClass {}
       ```

    c. ```
       /*This is a comment */

       package somepackage;
       import java.nio.*;
       class SomeClass {}
       ```

2. For a hierarchy of a resource property file, if a key is missing from one of the derived files, which of the following are true about the value returned, based on a missing key?

    a. The return value will be an empty string
    b. The return value will be a null value
    c. The return value will be a string from a base resource bundle
    d. A runtime exception will be thrown

3. Which exception is not thrown by the `forName` method:

    a. `InstantiationException`
    b. `ClassNotFoundException`
    c. `ClassDoesNotExistException`
    d. `IllegalAccessException`
       ```

A
Test Your Knowledge – Answers

Chapter 1: Getting Started with Java

Question No.	Answer	Explanation
1	a	The second command line argument is displayed.
2	a, b, and d	Option c is incorrect because you cannot assign a double value to an integer variable.

Chapter 2: Java Data Types and Their Usage

Question No.	Answer	Explanation
1	c	You cannot access an instance variable from a static method.
2	c and e	Option a is incorrect because single quotes are used for character data. Option b requires a suffix of f as in 3.14159f. A byte accepts only the values from -128 to +127.
3	b and d	Option a is incorrect because instance variables need to be used with an object. Option c is incorrect because instance variables cannot be used with a classname.
4	a, b, and d	There is no StringBuilder toInteger method.

Test Your Knowledge – Answers

Question No.	Answer	Explanation
5	b	The `lastIndexOf` method takes a single argument of char. The `charAt` method returns the letter at the position. The last use of `indexOf` does not take both a string and a char argument.
6	c	Option a only compares equality of objects. Option b is incorrect because there is no such method as `matchCase` In option d, the `equals` method uses the case which is different in the two strings.

Chapter 3: Decision Constructs

Question No.	Answer	Explanation
1	b	The `%` operator is the modulo operator and returns the remainder.
2	a and c	Option b evaluates to -24. Option d evaluates to 11.
3	b	The bit sequence 0001000 is shifted to the right 3 positions with a zero sign fill.
4	a and c	Option b results in a comparison between `i` and `j` which returns a Boolean value. This value cannot be compared against the integer `k`. Option d requires an operand before the expression, `>k`.
5	b	The default case can be positioned anywhere within the switch. As all of the cases, except the first one, are missing a break statement, flow falls through each of the last three cases. While it is not common, constants can be used for switch statements.

Appendix

Chapter 4: Using Arrays and Collections

Question No.	Answer	Explanation
1	a and d	The number of elements in an array declaration is not used in the declaration. However, we can use the following: `int arr[] = new int[5];`
2	b	At least the first dimension of a multidimensional array must be specified.
3	a and c	The `contains` method will return true if the object is found and `indexOf` takes an object reference and returns the index of the object if found, otherwise it returns a -1. The `indexOf` method does not take an integer argument and the hasObject method does not exist.

Chapter 5: Looping Constructs

Question No.	Answer	Explanation
1	a and d	The other options will not work because the expression does not evaluate to a Boolean value.
2	b, c, and d	You cannot use `[]` in a for-each statement.
3	a and b	Option c needs parentheses around the expression, `i < 5`. Option d requires a block statement if more than one statement is used between the `do` and `while` keywords.
4	a, b, c, and d	They are all equivalent.
5	a	The continue statement skips values 3 for `j`.

Chapter 6: Classes, Constructors, and Methods

Question No.	Answer	Explanation
1	a, c, and d	Option b fails to initialize the array properly.
2	c	You cannot access an instance method from a static method.
3	a	The return value is not considered when overloading a method.
4	c and d	The last line is a method that happens to have the same name as the constructor. As there are constructors defined but no default constructor, the class has no default constructor.
5	a and b	Only the `private` and `public` keywords can be used when declaring a class and the `private` keyword can only be used with inner classes.
6	c	As the classes are on the same package and all of the methods are visible except for the private method.
7	d	The `i` variable in the main is not modified, as it is passed by value. While the string is passed by reference, the local variable s was modified in the third method, not the one in the `main` method.

Chapter 7: Inheritance and Polymorphism

Question No.	Answer	Explanation
1	a and d	This results in `ClassC` being the "grandchild" of `ClassA`.
2	a and d	Overloading only occurs within the same class. It is not necessary to have a base class. A common implemented interface will also work for polymorphic behavior.
3	d	The other methods do not exist.
4	b	The others will generate syntax errors.
5	c	The first answer is used only as the first statement of a constructor. The second answer generates a syntax error. The fourth option results in unbounded recursion.
6	a	An abstract class does not have to have abstract methods and can extend other classes, whether they are abstract or not. It is common to find abstract classes that implement an interface.

Chapter 8: Handling Exceptions in an Application

Question No.	Answer	Explanation
1	c and d	Checked exceptions are those classes that extend the `Exception` class but not the `RuntimeException` class.
2	b and c	The derived most class should be caught first. The order of classes at the same hierarchy level is not important.
3	d	Checked exceptions should be handled. They may be handled using a try-catch block or may be re-thrown to another method in the call stack which is better suited to deal with the exception.
4	a and d	We can normally handle checked exceptions and they should be used.
5	a and d	`DivisionByZeroException` does not exist. No file operations are performed here.

Chapter 9: The Java Application

Question No.	Answer	Explanation
1	a and c	The package declaration must come before any other code. However, comments may appear anywhere within the code.
2	c	A string from a base resource bundle will be returned, if present.
3	c	This exception does not exist.

Index

Symbols

@Deprecated annotation 35
@Override annotation 13, 35, 227, 228
@SuppressWarnings annotation 35

A

abstract classes 230
abstract methods 230
Abstract Windowing Toolkit (AWT) 36
access modifiers
  about 25, 60
  package 60
  package scoped 25
  private 25, 60
  protected 25, 60
  public 25, 60
accessor methods 23, 211
activation record 41
add method 143
annotations 35
anotherMethod 252
append method 75
applets 12
args parameter 43
array 40
arraycopy method 134
ArrayList class 140
  about 143
  ArrayList object, sorting 148
  ArrayList object, traversing 146, 147
  characteristics 143
  creating 144
  elements, adding 144, 145
  elements, retrieving 145
  for-each statement, using with 162
  methods 148
ArrayList methods 148, 149
ArrayList object
  about 140
  sorting 148
  traversing 146, 147
array of objects 123, 124
arrays
  about 118
  array of objects 123
  Arrays class 140
  characteristics 141
  multidimensional arrays 124
  one-dimensional arrays 118
  techniques 127
Arrays class
  about 141
  asList method 140
  deepToString method 140
  java.util.Arrays class 140
  toString method 140
arrays, comparing
  about 130
  deepEquals method, using 132, 133
  element-by-element comparison 131
  equality operator, using 131
  equals method, using 132
  techniques 130
arrays, copying
  about 133
  Arrays.copyOf method, using 136
  Arrays.copyOfRange method, using 136
  clone method, using 137
  deep copy 134
  shallow copy 134

simple element-by-element copy 134
System.arraycopy method, using 134-136
techniques 134
arrays, passing 137
arrays, traversing
about 127
for-each statement, using 129
simple loops, using 128, 129
array techniques
about 127
arrays, comparing 130
arrays, copying 133
arrays, passing 137, 138
arrays, traversing 127, 128
command-line arguments, using 139
asList method 140
autoboxing 49

B

base class 219
base class constructor
calling 235-237
basic coding activities 112
BigDecimal class 47, 280
BigDecimal constructor 43
boolean 46
Boolean variables
about 81, 82
declaring 81
break keyword 101
break statement 113, 171
byte 46

C

casting 63
catch block 250
C/C++ 8
char 46
Character class
about 64, 66
methods 66
character literals 55
Character methods
about 66
isDigit 66
isLetter 66
isLetterOrDigit 66
isLowerCase 66
isSpace 66
isUpperCase 66
toLowerCase 66
toUpperCase 66
charAt method 72
CharSequence class 64
Charset class 64
checked exceptions 251
class
about 20, 186
data encapsulation 188, 189
instance variables, referencing 189
memory management 187
object, creating 186, 187
signature 190
ClassCastException exception 243
class diagram 9
classes and objects, managing
about 234
base class constructor, calling 235-237
Object class 241
objects, casting 242, 243
object type, determining 240
overridden method, accessing in base class 237-239
scope 243-245
super keyword 235
CLASSPATH environmental variable 34
class structure
about 19
access modifiers 25
classes 19
constructors 21
documentation 26
interfaces 21
methods 19, 22
objects 20, 21
packages 19
class variable 212
clone method 137
code organization
about 277, 278
directory/file organization, of packages 278
garbage collection 283

import statement 280
packages 278
Collator class 64
Collections Framework
  about 142
  ArrayList class 143
  collections, encapsulating 149, 150
  interfaces and classes 142
  iterators 142
  ListIterator interface 143
  reference link 142
command-line arguments
  using, arrays 139
comma operator, for statement 156
comments 27
compareTo method 70
computePay method 233
conditional operator
  about 80, 99
  ElseExpression 99
  essential form 99
  LogicalExpression 99
  ThenExpression 99
Constants 51, 56
constructor chaining 204
constructors
  about 21, 199
  characteristics 199
  default constructor 200, 201
  issues 205
  Java initialization sequence 206
  overloading 202, 203
  private constructors 204, 205
continue statement 172
control statements
  about 80
  conditional operator 99
  if statement 87
  switch statement 100
Control structure issues
  about 105
  comparing objects 111
  decision constructs issues 105, 106
  floating point number considerations 106, 107
  goto statement 112, 113
  three basic coding activities 112

control structures 80
copyOf method 136
copyOfRange method 136
createStatement method 293
Currency class 48
Customer class 17
Customer class, Java application 18
Customer constructor 43
CustomerDriver class 15 19

D

data encapsulation 10, 11, 188
data handling
  about 40
  access modifiers 60
  autoboxing 49
  constants 51
  data summary 61
  enumerations 51
  heap 41
  identifiers, initializing 49, 50
  instance and static data 58
  Java identifiers 40
  lifetime 59
  literals 51
  memory 40
  objects 40
  primitive data types 46
  scope 58
  stack 41
  variable, declaring 45
  wrapper classes 48
data structures 40
data summary 61
decision constructs issues 105, 106
deep copy 134
deepEquals method 132
deepToString method 140
default constructor 200, 201
derived class 219
directories, IDE file structure
  .classpath 31
  .project 31
  .settings 31
  bin 31
  src 31

directories, SDK file structure
  bin 31
  db 31
  demo 31
  inlcude 31
  jre 31
  sample 31
displayArray method 137
documentation
  about 26
  comments 27
  Javadocs, using 28
  Java naming conventions 28
doGet method 13
Do nothing comment 90
doPut method 13
double 47
do-while statement
  about 169
  using 170
downcasting 242

E

element-by-element comparison 131
elements
  adding, to ArrayList 144
  retrieving, in ArrayList 145
elements, main method
  args 24
  public 24
  static 24
  void 24
else if variation 91
Employee base class 221
Employee class
  about 186
  methods 190
end loop operation, for statement 155
endsWith method 70
Enterprise Java Beans (EJB) 14
enumeration-based switch statements 103, 104
Enumeration object 288
enumerations 51, 57, 58
equality operator 82
equalsIgnoreCase method 70

equals method 70 132
Exception class
  about 251
  Checked 251
  Unchecked 251
exception handling
  about 250
  mechanisms 250
exception handling techniques 252
exception types 251
executeQuery method 294
expressions
  building, operators used 61

F

FILE_EXISTS key 290
fill method 140
finalize method 44
final keyword
  about 57
  using, with classes 229
finally block 250
float 47
floating point 47
floating point number considerations
  about 106
  floating point numbers, comparing 109, 110
  rounding errors 110
  special floating point values 107-109
  strictfp keyword 111
floating point numbers
  comparing 109, 110
flow of control, Java application
  about 80
  control statement 80
for-each statement
  about 160
  advantages 160
  drawbacks 161
  Iterator interface, implementing 164, 165
  usage issues 165
  using 160
  using, with list 162
for loop variations, for statement 158
for statement
  about 154

comma operator 156
end loop operation 155
for loop variations 158, 159
initial operation 154
scope 157
terminal condition 154
using 155

G

garbage collection 277, 283, 285
getBundle method 288
getConnection method 293
getDate method 199
getInstance method 204
getKeys method 288
getObject method 291
getString method 288
getter methods 23
getWidth method 212
goto statement 112
Graphical User Interface (GUI) 36 112
Graphical User Interface (GUI)
    development 8

H

hasNext method 143, 164
hasPrevious method 143
heap 41, 119

I

IDE file structure
    about 31
    directories 31
identifiers
    initializing 49, 50
IEEE 754 Floating Point Arithmetic standard
    URL 106
if statement
    about 80, 87-90
    else if variation 91, 92
    empty statement 90
    graphical depiction diagram 87, 88
    nesting 90
    usage issues 92

immutable objects
    about 58, 198
    creating 198
    declaring 198, 199
import statement, Java application
    about 17, 280
    avoiding 280
    multiple classes, with same name 282
    static import statement 283
    using 280
    wildcard character, using 281
indexOf method 72
IndexOutOfBoundsException exception 135
infinite loop
    about 159, 175
    using 175, 176
inheritance
    @Override annotation 227, 228
    about 11, 220
    abstract classes 230
    abstract methods 230
    final keyword, using with classes 229
    methods, overriding 225-227
    protected keyword, using 223-225
    subclass, implementing 221-223
initialAge parameter 191
initial operation, for statement 154
instance
    versus, static data 58
instance method 214
instance variables 45, 212
instance variables, Java application 18
int 46
integer 47
integer-based switch statements 101-103
Integrated Development Environments
    (IDE) 30
interfaces 21
intern method 70
isDigit method 66
isLetter method 66
isLetterOrDigit method 66
isLowerCase method 66
isSpace method 66
isUpperCase method 66

[307]

Iterator interface
  implementing 164
  MyIterable class 164
  MyIterator class 164
iterators, Collections Framework
  about 142
  hasNext method 142
  next method 142
  remove method 142

J

Java
  about 8
  arrays 117, 118
  break statement 171
  characters 64
  classes 186
  Collection Framework 142
  conditional operator 99
  constructors 199
  continue statement 172
  data handling 40
  do-while statement 169
  enumerations 57, 58
  exception handling techniques 252
  final keyword 57
  for-each statement 160
  for statement 154
  if statement 87
  immutable objects 58
  infinite loops 175
  instance methods 214
  instance variable 212
  labels, using 174, 175
  logical expressions 81
  looping constructs 153
  methods 207
  nested loops 172
  object-oriented software development 8
  OOP principles 10
  programming constructs 79
  static method 214
  static variable 212
  StringBuffer class 74
  StringBuilder class 74
  strings 64
  this keyword 190
  types of Java applications 12
  Unicode characters 65
  while statement 167
Java 2 Enterprise Edition (J2EE) 31
Java application
  code organization 277
  compiling 30
  exceptions, handling 250
  flow of control 80
  JDBC 292
  Locale class 285
  resource bundles 287
Java application development process
  annotations 35
  IDE file structure 31
  investigating 29
  Java application, compiling 30
  Java class libraries 35, 36
  Java environment 33, 34
  on Windows platform, Java 7 used 32, 33
  SDK file structure 31
Java applications
  compiling, on Windows platform using Java 7 32
  executing, on Windows platform using Java 7 32
Java Archive (JAR) file 30, 279
java.awt 36
Java class libraries
  about 35
  java.awt 36
  java.io 36
  java.lang 36
  java.net 36
  java.util 36
Java console program structure
  exploring 15
  simple Java application 15-17
Java Development Kit (JDK) 8, 30
Javadoc tool
  about 28
  using 28
Java Enterprise Edition (JEE) 31
Java environment
  about 33
  variables 34

JAVA_HOME root directory 31
JAVA_HOME variable 34
Java initialization sequence 206
java.io 36
java.lang 36
Java naming conventions
  about 28
  rules and examples 28
java.net 36
Java Runtime Environment (JRE) 30
JavaServer Page (JSP) 13
java.util 36
java.util.ArrayList class 117
java.util.Arrays class 117
java.util.Collection interface 160
java.util.Iterable interface 160
JAVA_VERSION variable 34
Java Virtual Machine (JVM) 8, 29, 249
JDBC
  connection, establishing 293
  database, connecting to 292
  driver, loading 292
  results, handling 294, 295
  SQL statement, creating 293
  using 292

L

labels
  about 174
  using 174
lastIndexOf method 72
length method 72
lifetime 59
ListIterator interface
  about 143
  methods 143
ListResourceBundle class
  using 290, 291
literal 51
literal constants
  about 51
  numeric literals 51
Locale class
  about 18, 285
  using 286
local variables 45

logical expressions
  about 81
  Boolean variables 81
  equality operator 82
  logical operators 83, 84
  operands 81
  operators 81
  relational operators 82, 83
  short circuit evaluation 85
logical operators
  ! 84
  && 84
  || 84
  about 81, 83
  using 84
long 47
looping constructs
  about 153
  do-while statement 169
  for-each statement 160
  for statement 154
  pitfalls 179, 180, 181
  while statement 167

M

main method
  about 12, 24
  elements 24
  value, returning from application 25
memory management 187
methods
  about 22, 207
  accessor method 211
  calling 208
  declaring 22, 23
  defining 207
  main method 24
  mutator method 211
  overloading 209, 210
  overriding, in inheritance 225
  signature 23
methods, ListIterator interface
  add 143
  hasNext 143
  hasPrevious 143
  next 143

[309]

nextIndex 143
previous 143
previousIndex 143
remove 143
set 143
methods, Object class
   clone 242
   equals 242
   finalize 242
   getClass 242
   hashCode 242
   notify 242
   notifyAll 242
   toString 242
   wait 242
miscellaneous String methods
   about 74
   replace 74
   tolowercase 74
   toUpperCase 74
   trim 74
   using 74
multidimensional arrays 124-126
mutator methods 23, 211
MyIterable class 164
MyIterator class 164

N

narrowing 63
nested if statements 90
nested loops
   about 172
   using 172-174
nextIndex method 143
next method 143
NullPointerException 249
NullPointerException exception 135
number/string conversions 73
numeric literals
   about 51, 52
   using 53, 54

O

Oak 7
object
   creating 186, 187

Object class
   about 241, 242
   methods 242
Object Oriented Analysis. *See* OOA
Object Oriented Design. *See* OOD
Object Oriented Programming. *See* OOP
object-oriented (OO) language 8
Object-oriented software development 8
objects
   about 21
   casting 242
   comparing 111
object type
   determining 240, 241
one-dimensional arrays
   about 118-120
   ages array 119
   initializing 121, 122
   placement of brackets 120
OOA 9
OOD 9
OOP 9
OOP principles
   about 10
   data encapsulation 10
   inheritance 11
   polymorphism 11
OO technologies
   OOA 9
   OOD 9
   OOP 9
operands 61
operators
   classifying 62
OS_ARCH variable 34
OS_NAME variable 34
OS_VERSION variable 34
out variable 13
overloaded constructors 202, 203
overridden method
   accessing, in base class 237-239

P

package-private 243
packages, code organization
   about 278

[310]

classes 278
directory/file organization 278, 279
enumerations 278
exceptions 278
interfaces 278
package statement, Java application 17
passing by value 194
PATH variable 34
person class 219
placement of brackets 120, 121
polymorphism
   about 11, 220, 231
   using 232-234
precedence and associativity table 62, 63
PREFIXES key 291
previousIndex method 143
previous method 143
primitive data types
   about 46
   boolean 46
   byte 46
   char 46
   double 47
   float 47
   int 46
   long 47
   short 46
println methods 13
private constructors 204, 205
program errors 249
programming constructs
   about 79
   conditional operator 99
   control statement 80
   if statement 87
   logical expressions 81
   switch statement 100
property resource bundle
   using 287-290
protected keyword
   using, in inheritance 223, 224
public keyword 244

R

ragged arrays 126
read-only fields 212

read-only member variable 23
Rectangle class 212
reference variable 189
relational operators
   < 83
   <= 83
   == 83
   > 83
   >= 83
   about 82, 83
remove method 143, 148
replace method 74
residual value 93
resource bundles
   about 287
   ListResourceBundle class, using 290, 291
   property resource bundle, using 287-290
ResourceExamples_en_US.properties file 288
rounding errors 110
RuntimeException class 251

S

SalaryEmployee class 221
scope
   about 58, 243
   review 243
scope, of variables
   explaninig 60
scoping rules 59
SDK file structure
   about 31
   directories 31
servlet 13
setAge method 189
setBalance method 43
set method 143
setName method 22
setter methods 23
shallow copy 134
short 46
short circuit evaluation
   about 85
   avoiding 86
short circuiting
   about 85

logical && operator, using 85
logical || operator, using 86
signature 190
signature, method 24
simple element-by-element copy 134
simple Java application
  about 15
  ackage statement 17
  customer class 18
  CustomerDriver class 19
  import statement 17
  instance variables 18
  methods 18
SJava Software Development Kit (SDK) 31
someMethod 252
special floating point values 107-109
SQL statement
  creating 293
stack 41
stack frame 41
startsWith method 70
static import statement 283
static/instance variables and method
  relationship 214
static method 214
static setMinimumAge method 213
static variables 45, 212
StreamTokenizer class 64
strictfp keyword 111
string-based switch statements 104
StringBuffer class 64, 74
StringBuilder array 135
StringBuilder class 18, 64, 74
String class
  about 64-67
  miscellaneous methods 74
  Number/string conversions 73
  String comparisons 68-71
  String length 73
  String methods 72
string classes
  using 64, 65
String comparisons
  performing 68-71
  using 70, 71
String data type
  about 47

BigDecimal class 47
floating point 47
integer 47
string issues, switch statements 105
String length 73
string literals 56
String methods
  about 70
  charAt 72
  compareTo 70
  endsWith 70
  equals 70
  equalsIgnoreCase 70
  indexOf 72
  lastIndexOf 72
  length 72
  startsWith 70
  substring 72
String objects 67, 68
StringTokenizer class 64
subclass
  implementing, in inheritance 221
substring method 72
summation process 177-179
super keyword 235
supplementary characters 66
surrogate pairs 66
switch statement
  about 80, 100
  enumeration-based switch statements 103
  integer-based switch statements 101-103
  string-based switch statements 104
  string issues 105
  using 101

T

terminal condition, for statement 154
ternary operator 99
this keyword
  immutable objects 198
  parameters, passing 193, 195
  uses 190-192
  variable number of arguments 196, 197
thread 65
Throwable class 251
toLowerCase method 66, 74

toString method
  about 18, 122, 140
toUpperCase method 66, 74
trim method 74
try block 250
types, Java applications
  about 12
  Applets 12
  Enterprise Java Beans (EJB) 14
  JavaServer Page (JSP) 13
  servlet 13

U

unboxing 49
unchecked exceptions 251
Unicode characters 65, 66
Unified Modeling Language (UML) 9
UnsupportedOperationException exception 162
upcasting 242
usage issues, for-each statement
  about 165
  null values 166
  variable number of arguments 167
usage issues, if statement
  about 92
  block statement, avoiding 96, 97
  Boolean variables, using instead of logical expressions 94, 95
  dangling else problem 97, 98
  equality operator, misusing 92
  inverse operations, using 93, 94
  true and false keywords, using 95
UTF-16 (16-bit Unicode Transformation Format) 66

V

valid variable examples
  $newValue 45
  _byline 45
  _engineOn 45
  mileage 45
  numberCylinders 45
  numberWheels 45
  ownerName 45
variables
  declaring 45
  instance variables 45
  local variables 45
  static variables 45

W

while statement
  about 167
  using 168, 169
WINDOW_CAPTION key 290
wrapper classes
  about 48
  data types 48
write-only fields 212
write-only member variable 23

[PACKT] enterprise
PUBLISHING
professional expertise distilled

Thank you for buying
Oracle Certified Associate, Java SE 7 Programmer Study Guide

About Packt Publishing

Packt, pronounced 'packed', published its first book "Mastering phpMyAdmin for Effective MySQL Management" in April 2004 and subsequently continued to specialize in publishing highly focused books on specific technologies and solutions.

Our books and publications share the experiences of your fellow IT professionals in adapting and customizing today's systems, applications, and frameworks. Our solution based books give you the knowledge and power to customize the software and technologies you're using to get the job done. Packt books are more specific and less general than the IT books you have seen in the past. Our unique business model allows us to bring you more focused information, giving you more of what you need to know, and less of what you don't.

Packt is a modern, yet unique publishing company, which focuses on producing quality, cutting-edge books for communities of developers, administrators, and newbies alike. For more information, please visit our website: www.packtpub.com.

About Packt Enterprise

In 2010, Packt launched two new brands, Packt Enterprise and Packt Open Source, in order to continue its focus on specialization. This book is part of the Packt Enterprise brand, home to books published on enterprise software – software created by major vendors, including (but not limited to) IBM, Microsoft and Oracle, often for use in other corporations. Its titles will offer information relevant to a range of users of this software, including administrators, developers, architects, and end users.

Writing for Packt

We welcome all inquiries from people who are interested in authoring. Book proposals should be sent to author@packtpub.com. If your book idea is still at an early stage and you would like to discuss it first before writing a formal book proposal, contact us; one of our commissioning editors will get in touch with you.

We're not just looking for published authors; if you have strong technical skills but no writing experience, our experienced editors can help you develop a writing career, or simply get some additional reward for your expertise.

Java 7 New Features Cookbook

ISBN: 978-1-849685-62-7　　　Paperback: 384 pages

Over 100 comprehensive recipes to get you up-to-speed with all the exciting new features of Java 7

1. Comprehensive coverage of the new features of Java 7 organized around easy-to-follow recipes

2. Covers exciting features such as the try-with-resources block, the monitoring of directory events, asynchronous IO and new GUI enhancements, and more

3. A learn-by-example based approach that focuses on key concepts to provide the foundation to solve real world problems

Java 7 JAX-WS Web Services

ISBN: 978-1-849687-20-1　　　Paperback: 64 pages

A practical, focused mini book for creating Web Services in Java 7

1. Develop Java 7 JAX-WS web services using the NetBeans IDE and Oracle GlassFish server

2. End-to-end application which makes use of the new clientjar option in JAX-WS wsimport tool

3. Packed with ample screenshots and practical instructions

Please check **www.PacktPub.com** for information on our titles